RAILWAYS IN IRELAND

Part One: Great Northern, SL&NC, Lough Swilly, County Donegal, Cavan & Leitrim, Clogher Valley, C&VB

Martin Bairstow

No 85 "Merlin" pauses at Goraghwood with a Belfast to Dublin express on 7 September 1957. A 4 – 4 –2T waits at the head of the Warrenpoint connection. *(J C W Halliday)*

Published by Martin Bairstow, 53 Kirklees Drive, Farsley, Leeds
Printed by The Amadeus Press, Cleckheaton, West Yorkshire

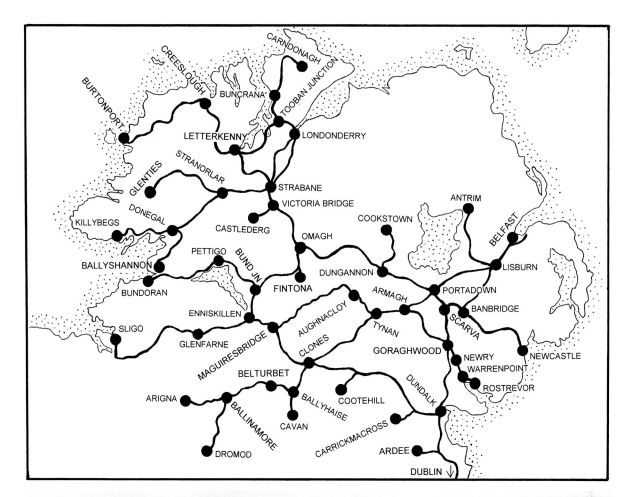

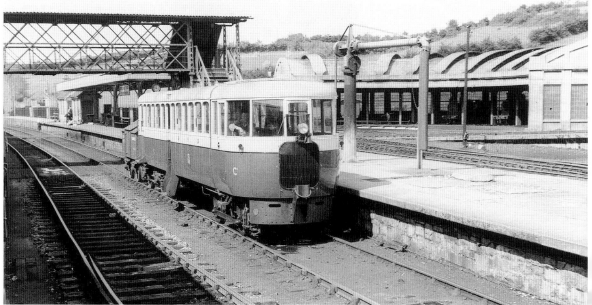

G N diesel railcar C1 with luggage trailer arriving at Clones from Cavan in 1956. There were only two half roundhouse engine sheds on the G N. Clones dated from 1926. A similar but larger one was built at Portadown, in the angle between the Dublin and Armagh lines.

(Peter Sunderland)

The Railways of Ireland
In Five Parts

This book is a new, slightly extended edition of the title which first appeared in 2006. At that time, I expressed the hope of covering in four parts all the public railways of Ireland, together with their associated steamship services. In the event, the work ran to five parts, divided by reference to the pre Grouping companies.

Part One
Great Northern; Sligo, Leitrim & Northern Counties; Londonderry & Lough Swilly; County Donegal Joint; Cavan & Leitrim; Clogher Valley; Castlederg & Victoria Bridge

Part Two
Belfast & County Down; Northern Counties; Giants Causeway; Bessbrook & Newry; Dundalk, Newry & Greenore; Ballycastle; Belfast Tramways; Cross Channel Shipping; Northern Ireland Railways.

Part Three
Dublin & South Eastern; Midland Great Western; Royal Mail Route; Fishguard & Rosslare; Dublin United Tramways; Dublin & Lucan; Dublin & Blessington; Galway & Salthill.

Part Four
Great Southern & Western

Part Five
Cork, Bandon & South Coast; Cork & Macroom; Timoleague & Courtmacsherry; Cork, Blackrock & Passage; Cork & Muskerry; Schull & Skibbereen; Cork Tramways; City of Cork Steam Packet Company; West Clare; Tralee & Dingle; Listowel & Ballybunion

Contents

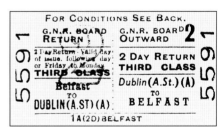

SG class 0 – 6 – 0 No 178 shunting at Ardee on 20 October 1959. The branch had closed to passengers in 1934 but remained in use for goods until 1975.
(R M Arnold)

The Great Northern Railway (Ireland)

4 – 4 – 0 No 170 "Errigal" at Omagh on a Londonderry to Belfast express in September 1957.

(J C W Halliday)

The "Ireland" qualification was necessary because there already was a Great Northern Railway in England. The Irish concern was created by amalgamation on 1 April 1876. It survived as an independent undertaking until 30 September 1958, its last five years under Government ownership.

The amalgamation of 1876 brought together four concerns:

The Dublin & Drogheda Railway
The Dublin & Belfast Junction Railway
(Drogheda to Portadown)
The Ulster Railway
(Belfast to Portadown, Clones and Omagh)
The Irish North Western Railway
(Dundalk – Clones – Omagh – Londonderry)

The first three made up the Dublin to Belfast Main Line

At its largest extent, the Great Northern covered 617 route miles. It also had a half share in the narrow gauge County Donegal Joint Committee. It was a pioneer in the use of diesel railcars. For 25 years up to 1957, it had the distinction of employing four types of motive power on its passenger service; steam, electric, diesel and horse.

Regarded by many as the most progressive Irish railway, the Great Northern suffered badly from the Border, which it crossed in 15 places. It fell foul of the anti rail policy in Northern Ireland and lost proportionately more of its mileage than the other principal railways. Very little survives apart from the Dublin to Belfast Main Line.

This outline history begins by looking at the four constituent companies. The story is not simple. Some branches were only leased or worked by the larger companies. Others were built after the amalgamation. Some are best grouped geographically.

The Ulster Railway

The oldest constituent was the Ulster, incorporated in May 1836 with powers to build a line between Belfast and Armagh. About the same time, the Government set up a Royal Commission "to consider and recommend a General System of Railways for Ireland". The Ulster Company asked the Commission to predict, in advance of their Report, what was going to be their recommendation on the question of gauge. The answer given was 6ft 2in. This was accepted and adopted by the Ulster Railway. A single line opened from Belfast to Lisburn on 12 August 1839. Progress towards Armagh was impeded by difficulty in raising capital, forcing the Company to seek a loan from the Irish Board of Works. With the aid of this, the line reached Lurgan on 18 November 1841 and Seagoe, a temporary terminus one mile short of Portadown, on 31 January 1842. Problems with foundations on soft ground delayed entry into Portadown until 12 September the same year. Powers for Armagh were allowed to lapse for the time being.

At the very beginning, only first and second class carriages were conveyed, respectively with upholstered and wooden seats. This ignored the huge potential of working class travel into Belfast. From March 1840, open wagons were provided for what one of the directors described as "the great unwashed". Over the next few years, the Company wavered between attaching open thirds to passenger trains or sending them with the goods. It was not until 1856 that all third class passengers were accommodated in covered carriages fitted with seats.

The matter of gauge was reopened in 1843 when the Dublin & Drogheda Railway adopted 5ft 2in. This line was intended eventually to meet with the Ulster in order to form a Main Line between Dublin and Belfast. The Ulster Railway, believing that 6ft 2in was the official Irish gauge, asked the Board of Trade to compel the Dublin & Drogheda to follow the recommendation of the 1836 Royal Commission. The Dublin & Drogheda countered that the Ulster should be made to convert.

The Board of Trade asked its Inspector General of Railways, Major General Pasley, to arbitrate. He canvassed the views of a number of engineers whose advice varied from 5ft to 5ft 6in. He evidently didn't include Brunel in his survey. For want of a better idea, he averaged 5ft and 5ft 6in, neither of which was in existing use, and came up with 5ft 3in, which had not yet been adopted either. This became the Irish standard under the Regulation of Railways Act, 1846. The same legislation confirmed the gauge of 4ft 8½ in Great Britain, outside of Brunel's 7ft Great Western empire. Nobody considered any need

to make the two islands the same, as they could foresee neither the train ferry nor the economics of manufacturing modern rolling stock.

Pasley's ruling fell much harder on the Ulster Railway, already operating on 6ft 2in, than on the unfinished Dublin & Drogheda which merely had to broaden by one inch. The 1846 Act did not compel pre-existing railways to convert but, in practice, it left them with little choice. The 1845 Act for the revived Armagh project had specified 5ft 3in and all future extensions and connections would have to be at the new standard. The Ulster Railway accepted the inevitable and laid a second track, of 5ft 3in, between Belfast & Portadown. It then narrowed the original single line so as to create double track. The formation was wide enough because double track had always been the ultimate objective.

Passenger operation on 5ft 3in gauge began on 4 January 1847. For a short period, goods traffic was worked along the parallel 6ft 2in gauge track until this was taken out of use about May 1847. The broader gauge had lasted less than eight years. The Company was entitled, by statute, to contributions towards the cost of regauging, from four connecting railways, some still in the formative stage. It took until 1856 to collect all the money, some of which had to be sued for.

During its independent existence, the Ulster Railway owned a total of 68 locomotives, not all at once. Most of the early ones were built in Manchester by Sharp Roberts, later Sharp Brothers, later still Sharp Stewart. Some of the original 6ft 2in gauge engines were adapted to 5ft 3in. From 1864, the Railway was able to carry out rebuilds in its own Works in Belfast. From 1870, most new engines were built there, though some were still contracted out to Sharp Stewart or to Beyer Peacock, also of Manchester.

The most popular wheel arrangements were 2 – 2 – 2 for passenger engines and 0 – 4 – 2 for goods. Eight 0 – 6 – 0s, built from 1872, enjoyed a long life. They were withdrawn between 1924 and 1948.

Almost brand new Class S No 172 "Slieve Donard" leaving Dublin Amiens Street with an express for Belfast on 22 July 1914.
(LCGB - Ken Nunn Collection)

0 – 6 – 0 No 178 passing Howth Junction with a southbound goods in August 1957. The loco was supplied by Beyer Peacock in 1913 and withdrawn by CIE in 1961. *(Peter Griffin)*

The ultimate in Great Northern steam power were the five VS class 4 – 4 – 0s produced by Beyer Peacock in 1948. No 209 "Foyle" heads a Belfast to Dublin express on that same August Saturday. *(Peter Griffin)*

Railcar E coming off the Howth branch. Dating from 1936, this was one of four diesel trains, very advanced for their time. On the break-up of the GN in 1958, E went to CIE but was later sold to the UTA who cannibalised it to keep the other three going until 1965, when line closures brought about their scrapping. *(Peter Griffin)*

Ulster locomotives carried names, implying speed for passengers, with "Express", "Vulcan", "Fury" and "Spitfire", and strength for goods with "Samson" and "Hercules".

The Dublin & Drogheda Railway

This enterprise started with a very expensive Parliamentary battle against an alternative Dublin – Navan – Armagh scheme. The argument was whether the eventual Dublin to Belfast main line should follow a coastal or an inland route. The Dublin & Drogheda Act was passed in August 1836 but that was not the end of the matter.

In July 1838, just as construction work was getting started, the Royal Commission on Irish Railways issued a recommendation in favour of the Inland route. The Dublin & Drogheda Company immediately stopped work and lobbied the Government against a policy, which they feared would prevent their line from ever becoming the principal route to the North. They need not have feared. It quickly became clear that the Government had no plans to implement the Commission's proposal. But time had been wasted and the Dublin & Drogheda had to seek a renewal of its powers in an Act of 1840. Work was then able to resume in October.

All effort was concentrated on getting the line ready for traffic and, until this was achieved, nothing was spent on the terminal in Dublin. On 24 May 1844, the ceremonial opening of the Railway was combined with laying the foundation stone at Amiens Street Station. This took two years to complete. Until then, a temporary facility was used.

The line left Amiens Street by a 75 arch viaduct, which led on to the bridge over the Royal Canal. This was a single span, exceptionally long by contemporary standards. The alternative would have been to place an intermediate pier in the waterway but the Canal Company had demanded too high a price for this. In 1862, it became necessary to sink two support piers into the bed of the Canal. By that time, the waterway was owned by the Midland Great Western Railway. They took a more sympathetic view towards accommodating their neighbour.

In three places, the Dublin & Drogheda encountered broad river estuaries. Each was crossed by a stone faced embankment with a timber viaduct over the river itself. At Clontarf, the land on the inland side of the line was subsequently reclaimed. If this was the intention also at Malahide and Rogerstown, it never happened. The train still passes with water on both sides. The 572ft wooden viaduct at Malahide was the first to be replaced with

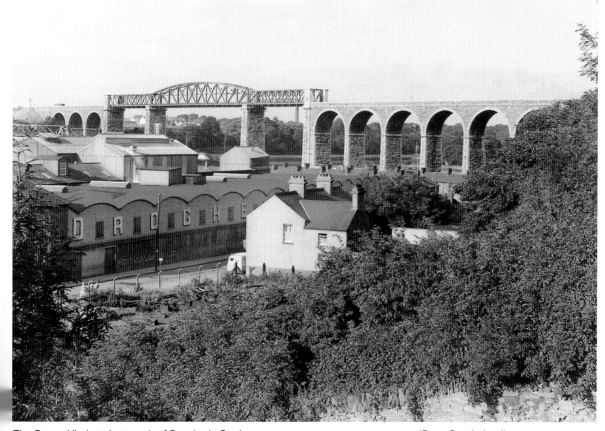

The Boyne Viaduct, just north of Drogheda Station.

(Peter Sunderland)

A Dublin to Belfast DMU crossing Craigmore Viaduct in July 1957. The path in the foreground is the track bed of the Bessbrook & Newry Tramway, closed in 1948. The view is towards Newry.

(Peter Sunderland)

stone piers and iron girders in 1860. The others followed. Further north, Balbriggan Harbour was crossed by an 11 arch viaduct, built in stone from the outset.

At Drogheda, a temporary wooden station sufficed for the first nine years, pending extension northwards over the Boyne Viaduct. The elevated position of the Railway at Drogheda prevented a connection ever being made with the port.

The Dublin & Drogheda also looked to Sharp Brothers for its early locomotives. After 1858, the majority came from Beyer Peacock. A total of 34 locos were built, of which 21 survived into GN ownership. The most popular wheel arrangement was 2 – 2 – 2.

The Dublin & Belfast Junction Railway
Incorporated in July 1845, the purpose of this line was to close the 63 mile gap between Drogheda and Portadown.

Setting aside the enormous task of bridging the Boyne at Drogheda, the route as far as Dundalk was relatively easy. This opened on 15 February 1849 from a temporary station at Newfoundwell, ¾ mile north of Drogheda Station. Transfer of goods and passengers between the two was very difficult because everything had to be carried down to river level, and then back up again.

The Company was having difficulty raising capital and had to turn to the Government for a loan. This was granted on condition that priority should be given to crossing the Boyne. But before contemplating this huge undertaking, they pressed ahead on the mountainous stretch north of Dundalk. Starting work from both ends, they opened 10½ miles from Dundalk to the Wellington Inn on 31 July 1850, then from Portadown to Mullaghglass on 6 January 1852. The last and most difficult six miles, including the 18 arch Bessbrook Viaduct, were completed on 10 June 1852.

The Boyne Viaduct
Attention then turned to crossing the River Boyne. This was achieved with a timber viaduct over which trains could pass at very low speed from 22 June 1853. This temporary structure also served as scaffolding for men working on the permanent stone and iron viaduct.

The iron girder portion comprised a central span of 226ft, flanked by two spans of 141ft each. Leading into these were 12 stone arches on the south side and three to the north. Considerable problems were experienced in the search for secure foundations, needed for the four heavy piers supporting the iron spans. The contractor, William Evans of Cambridge, had grossly underestimated the task. He became

The entrance to Belfast Great Victoria Street in 1955. *(Peter Sunderland)*

Class Q 4 – 4 – 0 No122 at Belfast Great Victoria Street in 1955, working the 5pm to Londonderry and Enniskillen. *(Peter Sunderland)*

insolvent and the work had to be finished with direct labour at double the original contract price. The permanent viaduct opened on 5 April 1855. With this, the Dublin to Belfast line was complete.

The viaduct was built for double track but, after 30 years of use, the signalling was adapted to prevent trains passing over in both directions at the same time. After 1920, the structure was subject to a weight restriction which precluded its use by some of the newest and most powerful engines. Reconstruction took place between 1930 and 1932, when the iron spans were replaced by narrower steel girders. Traffic continued during the work which left the bridge wide enough for only one track but with a higher permitted axle load.

Motive Power

In ordering its first locos, the Dublin & Belfast Junction followed the practice of its two neighbours and put its faith in Sharp Brothers 2 – 2 – 2s. The original intention was to have 16 but only nine were built over the period 1848 to 1854. The rest of the order was cancelled. Instead, the Company acquired six 0 - 4 – 2 goods engines from the same supplier.

Later, they became a customer of Beyer Peacock who supplied two 0 – 6 – 0s in 1872. As GN Nos 40 and 41, these survived until 1937 and 1934 respectively. The Dublin & Belfast Junction did not name its engines. Its locomotive headquarters were at Dundalk, a relatively small facility near Square Crossing, which never attempted to build new engines.

Train Services on the Dublin to Belfast Main Line

Pre 1914, the fastest trains in Ireland were those connecting with the overnight cross channel steamers. The shortest journey time between Dublin and Belfast was by the 6.05am from Amiens Street, the "Limited Mail" also advertised as a " Breakfast Saloon Train" with through carriages from Kingstown. With stops only at Drogheda, Dundalk and Portadown, this reached Belfast Great Victoria Street at 8.40. The corresponding up working was the 5pm from Belfast which, with the same stops, reached Amiens Street at 7.35, and Kingstown at 7.57pm.

The closest rival in terms of speed was the 7.30am "Breakfast Saloon Express" from Belfast, which reached Dublin in exactly three hours, with

No 210 "Erne" ready to leave Dublin Amiens Street with the 6pm Sunday express to Belfast in August 1956.   *(Peter Sunderland)*

stops at Portadown, Scarva, Goraghwood, Bessbrook, Adavoyle, Dundalk and Drogheda. The return was at 5.50pm from Dublin with stops at Balbriggan, Laytown, Drogheda, Dundalk, Goraghwood, Scarva, and Portadown. This reached Belfast in 3 hours 10 minutes. It slipped coaches at Malahide and Skerries, in the latter case for first and second class season ticket holders only. There were two other expresses at 9am and 3pm from Dublin, 9.55am and 2.20 from Belfast covering the 112½ miles in around 3¼ hours.

Anyone finding it necessary to be in Dublin early could leave Belfast by the 10pm, which included sleeping cars, arriving in Dublin at 5.10am. This train also picked up sleeping cars from Londonderry at Portadown. The corresponding working left Dublin at 8.20pm, with sleeping cars for Londonderry, but the Belfast portion arrived Great Victoria Street 10 minutes after midnight making sleeping accommodation unnecessary. This pair of evening/overnight trains ran seven days a week. Otherwise, the Sunday service was rather sparse. There was just one daytime through train, at 4.45pm from Dublin and 2.30pm from Belfast, taking 3½ hours.

The first three stations out of Dublin, Clontarf, Raheny and Junction, benefited from the Howth branch trains (see later). But north of Junction (later Howth Junction) there was no measurable suburban service into Dublin. The first train in from Skerries reached Dublin at 8.50am and may have been aimed at "white collar" commuters. There was nothing for early morning workmen, at least not in the up direction. There was a 6.25am to Skerries and a 6.50am first stop Skerries, which continued all stations to Lurgan then Lisburn and Belfast. These trains may have been more important for mail and parcels than for passengers. Most of the wayside stations enjoyed at least five trains in each direction, except Sundays when some had only one. Further north, the service was enhanced by trains coming off the Newry line at Goraghwood and from

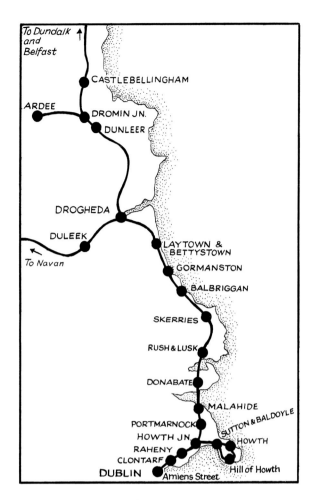

the Clones and Omagh routes at Portadown.

The 1938 timetable shows the gains achieved by the class V 4 – 4 –0s, following rebuilding of the Boyne Viaduct. The "Limited Mail" now leaves Dublin at 6.40am and takes exactly 2½ hours with

Portmarnock Station looking north in GN days. Since 1999, the Main Line has been electrified as far as Malahide.
(Geoffrey Lewthwaite)

Class S No 170 "Errigal" arriving at Drogheda in August 1959, with a Sunday afternoon slow from Dublin. CIE did not alter the running numbers of GN locos. *(Peter Sunderland)*

No 121 passing Bessbrook in July 1957, with a slow from Dundalk to Belfast. Closed in 1942, this station reopened in 1984, adopting the name Newry. *(Peter Sunderland)*

stops at Drogheda, Dundalk, Goraghwood, Portadown and Lurgan. The departure time appears to be 35 minutes later than in 1910 but 25 minutes of this comes from Ireland adopting London rather than Dublin time, an important consideration for a train with through carriages from Dun Laoghaire (formerly Kingstown). Up until 1916, the Irish Railways had set their clocks and watches by the time signal which rang down the telegraph wires at 10am Greenwich Mean Time, 9.35 Dublin Time.

In 1938, there were four expresses, besides the "Limited Mail", connecting Dublin and Belfast in 2½ to 3 hours, all with catering facilities and all having to stop at both Dundalk and Goraghwood for Customs examination. Up to 13 minutes were allowed at the latter place, which might not have justified a stop at all, had it not become a Customs post.

The Second World War put an end to the fast running until the "Enterprise Express" was introduced in 1948. The 1954 timetable shows six trains between Dublin and Belfast. There is no "Limited Mail". The 7.30am, 9.00, 2.45pm and 6.25 take from 3 hours 5 minutes to 3 hours 20, with up to 20 minutes standing at Goraghwood. But the 11am and 5.30pm are the "Enterprise Express" non stop in 2¼ hours. Customs examination was carried out at the termini. In return for this concession, the GN had to promise that the trains really would run non stop. Timetables were planned so as to minimise the risk of signal stops. If a conflict arose, signalmen had to give priority to keeping the "Enterprise" moving. If a driver needed to stop en route, he had, if at all possible, to do this only at Dundalk or Goraghwood, where Customs men could be summoned to attend by a whistle code to a strategic signal box. All this

4 – 4 –2T No 116 at Portadown with a local for Belfast on 8 May 1958. This is the third Portadown Station, east of the River Bann, which served passengers from 1863 until 1970. *(J C W Halliday)*

U.T.A.
1 DAY RETURN
SECOND CLASS
Moira
TO
BELFAST (G.V.)
DITIONS SEE BACK
Belfast.
2389

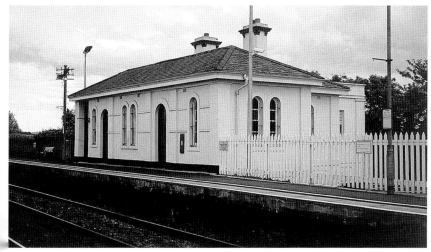

The oldest station building in Northern Ireland is on the Dublin bound platform at Moira. After destaffing in 1969, it became derelict but was restored in 1991 by the Northern Ireland Environment & Heritage Service. It is now used by them as a workshop. *(Alan Young)*

G class 0 – 6 – 0 No 179 calls at Lisburn with a southbound local on 6 May 1958. *(J C W Halliday)*

Railmotor No 4 leaving Great Victoria Street for Lisburn on 17 September 1909.
(LCGB Ken Nunn collection)

was to prevent smuggling dutiable goods. It was nothing to do with the movement of people, which was unrestricted.

Suburban traffic built up between Belfast and Lisburn. In 1905 three railmotors were introduced, comprising a passenger coach articulated onto a 0 – 4 – 0 engine supplied from Glasgow by the North British Locomotive Company, a successor of Sharp Stewart. The rail motors provided an almost half hourly stopping service between Belfast and Lisburn. This was over and above the longer distance trains, which generally ran non stop or semi fast over this initial section out of Belfast.

Additional stations were opened in 1907 at Finaghy, Derriaghy and Hilden. Adelaide had opened in 1898. The railmotors were withdrawn in 1913 and replaced by push – pull trains.

Armagh, Clones and Cavan

Powers for the Ulster Railway to build the 10½ miles from Portadown to Armagh were revived in 1845. The greatest obstacle was that of bridging the River

Bann just south of Portadown Station. This was achieved with a timber structure of five spans, which lasted until replaced by a stone and iron viaduct in 1871.

The Armagh extension opened on 1 March 1848, thus fulfilling the original ambition of the Ulster Railway to connect Belfast and Armagh. The line was double track throughout, until it was singled between Richhill and Armagh in 1934.

The Ulster Company waited until 1854 before depositing a Bill for the 16½ mile extension to Monaghan. Royal Assent was given in June 1855 and a contract let to William Dargan early the following year. The line, always single, opened on 25 May 1858.

The next advance, of 12 miles to Clones, was authorised in 1856 and completed on 2 March 1863. This section was double track until 1932. Just outside Clones Station, the Ulster Railway met the Dundalk & Enniskillen, soon to be renamed the Irish North Western. Clones marked the end of the Ulster's undisputed territory, yet there remained a

Armagh Station, looking towards Clones in August 1957. The site is now an Ulsterbus garage. The large goods shed is still in existence. *(Peter Sunderland)*

Class PP 4 – 4 – 0 No 75 giving up the single line token as it enters Armagh with a train from Clones in August 1957. It is taking the right hand line so as to be nearer the main station building.

(Peter Sunderland)

The entrance to Armagh Station, which was demolished in 1963. Monaghan, which survives in non railway use, is in similar style. So was the 1863 to 1970 station at Portadown. *(J C W Halliday)*

further traffic objective at Cavan.

By an Act of August 1859, the 15½ mile Clones & Cavan Extension Railway was authorised as an independent Company, to which the four principal neighbouring railways would subscribe capital. The Ulster, Dublin & Drogheda, Dublin & Belfast Junction and Dundalk & Enniskillen were the joint owners. The last named was charged with the task of building the line. It opened on 1 April 1862, eleven months before the Ulster Railway itself reached Clones. It was single track.

At Cavan, the line from Clones made an end on junction with the Midland Great Western Railway, which had arrived there from Inny Junction, near Mullingar in 1856. All trains used the MGWR station.

In 1910, there were four passenger trains each weekday leaving Cavan for Belfast. Typical journey time for the 79¾ miles was about three hours. As you got nearer to Belfast, the service intensified with two more trains starting at Clones and two more from Armagh. On Sundays, there was nothing from Cavan but two trains started at Clones and one at Armagh.

Redhills Station, looking towards Clones in 1955.
(H C Casserley)

Cavan Station looking north in 1958. The CIE Metrovic diesel is working the goods on the line from Inny Junction, closed to passengers in 1947. In the distance, railcar C2 is backing into the GN platform, having visited the turntable.
(J C W Halliday)

The Belturbet Branch

The 4¼ mile branch to Belturbet was authorised on 7 July 1862, by the same Act as that which changed the name of the D & E to the Irish North Western. No immediate progress was made and the branch was not opened until 29 June 1885, well into the Great Northern era. Two years later, Belturbet became an interchange with the narrow gauge Cavan & Leitrim Railway. The 1910 timetable shows seven trains leaving Belturbet, Sundays excepted. Four of these gave smart connections at Ballyhaise into the Belfast trains. The others connected for Cavan.

By the late 1940s, there were only two trains on the branch, leaving Belturbet at 11.55am and 5.20pm. Both had connections to Clones, but there was a wait there if you wanted to go further. The 5.20 was scheduled as a "mixed", passenger and goods working. The branch train was based at Belturbet, where there was a small engine shed. The passenger service was withdrawn as part of the "Great Closure" of 1957. Goods traffic continued for 18 months until the end of March 1959, when the branch closed at the same time as the connecting Cavan & Leitrim Railway.

LQG class 0 – 6 – 0 No 158 at Ballyhaise on 2 August 1958. Built in 1906 and originally named "Ballybay", No 158 was one of the most powerful 0 – 6 – 0s in Ireland. Ballyhaise had three platforms with an island connected by footbridge to the main one adjoining the station building. Today this is well preserved as a private dwelling. The view is towards Cavan.
(J C W Halliday)

The transit shed at Belturbet, where goods were manhandled between the narrow gauge (left) and the GNR. The building on the extreme left is the overall roof of the passenger station.
(Peter Sunderland)

The Oldcastle Branch

Powers for the 16¾ mile branch from Drogheda to Navan were included in the Act incorporating the Dublin & Belfast Junction Railway in 1845.

A second Act transferred the proposed line to the Dublin & Drogheda Railway and authorised an extension to Kells. The Drogheda to Navan section opened on 15 February 1850. The extension to Kells followed on 11 June 1853.

The branch reached its ultimate terminus at Oldcastle on 17 May 1863. It was worked by two 2 – 2 – 2 Sharpe Stewart locos, which became No 17 "Apollo" and 18 "Diana" in GN stock. Oldcastle

generally had three trains a day taking 90 to 95 minutes for the 39 miles. There was a shed with space for two engines at Oldcastle. Railcar working speeded up the service to 80 minutes despite additional halts.

The passenger service was withdrawn on 12 April 1958. Complete closure beyond Navan Junction took place at the end of March 1963. General freight to Navan ceased in 1977, but the same year saw the start of zinc traffic from Tara Mines, ¼ mile towards Oldcastle. This and Platin Cement, two miles from Drogheda, keep the line open.

Class QG No 154 waiting to leave Navan for Drogheda on 2 June 1951. Navan Junction Station was ¼ mile beyond. The population of Navan has risen significantly from 4,000, in the 1920s, to a projected 60,000 in 2011. The Government plans an eventual reopening of the abandoned MGW route from Clonsilla but will not consider a passenger service over the freight only line from Drogheda. *(Desmond Coakham)*

Railbus No1 at Oldcastle in 1958. Built as "E" in 1934, it reopened the Scarva to Banbridge branch. It became E2 in 1936 and 1 in 1947. It covered 374,000 miles in 24 years of passenger service, then became a pw vehicle for a short time. It is preserved at Cultra. It was fitted with a reverse gear box so it could run equally well in either direction. There were turntables at both Drogheda and Oldcastle but the vehicle would always have to run backwards between the branch junction and Drogheda Station. *(Desmond Coakham)*

The Ardee Branch

Having successfully opposed an independent Bill in 1884, the Great Northern itself obtained powers for the five mile branch in 1892. The line opened on I August 1896. Gradients were minimal. There were no intermediate stations but one was provided at Dromin Junction, where the branch joined the Dublin to Belfast Main Line.

In 1910, the branch had a creditable seven trains each way, timed as far as possible to connect with trains on the Main Line. The passenger service ended in 1934 but goods traffic continued until November 1975. Dromin Junction station closed at the end of January 1955.

The Dublin & Antrim Junction Railway

Knockmore became a three way junction with the opening, on 13 November 1871, of the 18½ mile single track Dublin & Antrim Junction Railway. This had been authorised by an Act of 11 July 1861. As the title suggests, it was intended as a through route, connecting the Main Line from Dublin with the Belfast & Northern Counties Railway to Ballymena and beyond.

The Company experienced difficulty in raising finance and had to resort to a Board of Works loan. They did not attempt to operate their own line but leased it to the Ulster Railway for a period of seven years. The Antrim Company defaulted on the Board of Works loan. When the lease expired in November 1878, the line belonged to the Board of Works, as mortgagees in possession. They extended the lease temporarily, pending offering the line for sale by auction in January 1879. The Great Northern, as successor to the Ulster Railway, had its bid accepted and so assumed ownership of the branch.

In 1910, there were four trains each way and one on Sundays. They ran through from Belfast, taking about one hour to Antrim. There was a faster and more frequent service by the NCC, so the GN branch catered only for intermediate traffic. It was, however, a vital link for freight between the GN and NCC systems, which were not connected in Belfast except by the Harbour Commissioners' line with its severe curves and restricted clearances.

Expansion of Aldergrove Airport, during the Second World War, brought additional traffic to the branch. In 1942 a 2¼ mile branch was built to an aircraft factory at Gortnagallon, on the shore of Lough Neagh. Diverging 1 mile north of Crumlin, it catered for both goods and workmen but was dispensed with in 1947.

The 1949 timetable shows six trains through trains each way plus an extra from Aldergrove to Antrim at 5.30pm. This was an NCC train, which worked out empty to Aldergrove. In later years, it ran coupled to the back of 5.15 Antrim – Belfast with the NCC engine in the rear so it could depart from Aldergrove without running round. The branch passenger service was withdrawn in 1960. The line remained in use for freight and reopened to passengers in 1974 as part of a development which will be described in *Part Two*.

Class WT 2 – 6 – 4T No 2 approaching Antrim with the Aldergrove to Ballymena workmen`s train, formed of two vintage NCC bogie coaches in August 1959. *(Peter Sunderland)*

UTA 0 – 6 – 0 No 38 (GNR class SG No 16) approaching Antrim with an "up" local in August 1959. In GN days, it was "up" to Antrim because it was "down" to Belfast. In 1974, Northern Ireland Railways changed it so that it was "up" to Belfast. It is normally "up" to the headquarters of the Railway: Dublin for the GN, Belfast for NIR.   *(Peter Sunderland)*

UTA No 26 "Lough Melvin", formerly of the Sligo, Leitrim & Northern Counties Railway, with a special at Crumlin in May 1960.   *(Peter Sunderland)*

Between 1933 and 1939, the GN opened four halts on the Antrim branch. Two were between Brookmount and Ballinderry at Brookhill and Meeting House and one at Legatiriff between Ballinderry and Glenavy.

The other was at Millar`s Bridge, 2½ miles north of Aldergrove. Facilities were very limited, indeed non-existent but it lasted from 1938 until September 1960. The view is looking back towards Lisburn.
(J C W Halliday)

The Irish North Western Railway

4 – 4 – 0 No 74 calling at Enniskillen with a train from Dundalk to Omagh on 7 September 1957.

(J C W Halliday)

This enterprise began as the Dundalk & Enniskillen Railway, incorporated in July 1845. At first, the Company was authorised to build only as far as Clones. Powers thence to Enniskillen were vested in the neighbouring Newry & Enniskillen Railway but transferred when the latter Company proved unable to fulfil its ambitions.

Construction began in October 1845. The line opened as far as Castleblaney on 15 February 1849 but further progress was held back by delays in raising capital. Until the end of 1850, operation was subcontracted to William Dargan, who employed two 2 – 2 – 2s for passenger work and two 0 – 4 – 2s for the goods. All four had been built by T Grendon & Co, Drogheda who themselves took over the operation of the Railway for a short time after Dargan.

Ballybay was reached on 17 July 1854 and Newbliss a year later. Parliamentary powers were then obtained for a branch to Cavan but this only ever reached Cootehill, opened on 18 October 1860.

The main line pressed on, through Clones, to Lisnaskea, opened on 7 July 1858, Lisbellaw, 16 August 1858 and Enniskillen, which was reached on 15 February 1859. In anticipation of opening to Enniskillen, the Company ordered three 2 – 2 – 2 tender engines from Beyer Peacock. These became Great Northern Nos 53, 55 and 56, lasting until 1907 – 1914.

At Dundalk, the Enniskillen Company had its own terminus at Barrack Street, which it reached by a sweep round the edge of the town, having first crossed the Dublin & Belfast Junction Railway on the level at Square Crossing. Connection between the two lines was achieved by a west to north curve, which brought Enniskillen traffic into the Dublin & Belfast Station. This stood on the junction itself, with separate platforms for the two Railways. It was replaced in 1893 by a larger structure, ¼ mile to the north.

The first passenger service between Dundalk and Enniskillen comprised four trains each way on weekdays. These started from Barrack Street, proceeded to the West Junction and then set back into the Dublin & Belfast Station.

The Londonderry & Enniskillen Railway

At Enniskillen, the Dundalk line made end on connection with the line from Londonderry. This had been authorised at the same time, in July 1845 but had been completed more quickly.

The route had first been surveyed by George Stephenson in 1837. The 1845 Act was based on work supposedly performed by Sir John MacNeill, Professor of Engineering at Trinity College, Dublin. There was so much railway work going on that eminent engineers just lent their names to projects without attending to them in person. The Company turned to Robert Stephenson to have the surveys amended.

Construction began at the northern end of the line in October 1845. The first 14 miles, from Londonderry to Strabane, followed the west bank of the Foyle and were almost level. This section opened on 19 April 1847. For the first three years, the Londonderry terminus was some ¾ mile south of the Carlisle Bridge. The line was extended, in April 1850, to Foyle Road Station, which stood just north of the Bridge joining the two halves of the City. At this time, the Londonderry to Strabane line was physically isolated from any other railway, a situation not remedied until the link up at Enniskillen in 1859. However, from 1852, Londonderry had a second railway at Waterside Station on what was to become the Belfast & Northern Counties Railway. This was built on the other side of the River, directly opposite Foyle Road. In January 1868, the two termini were linked by a track belonging to the Harbour Commissioners, which passed along the lower deck of the Carlisle Bridge. Access to the bridge, from the lines on either side was via turntables. Goods wagons were winched across by capstans or drawn by horses. Engines and passenger coaches could not go across.

Construction south from Strabane depended on raising capital in the difficult years following the "Railway Mania" of the mid 1840s. After a short delay, progress became rapid and the line was extended southwards in five stages:

Strabane to Newtown Stewart on 9 May 1852
Newtown Stewart to Omagh on 13 September 1852
Omagh to Fintona on 15 June 1853
Fintona Junction to Dromore Road on 16 January 1854
Dromore Road to Enniskillen on 19 August 1854

As an isolated undertaking, the Londonderry & Enniskillen struggled to gain sufficient traffic. It built up a fleet of 15 locomotives but, as soon as it became connected to the Dundalk & Enniskillen in 1859, it negotiated to lease itself to the latter Company. The lease took effect from the beginning of 1860. After two years, the Dundalk & Enniskillen decided that its name was insufficient to denote a line stretching from coast to coast. It re-christened itself the Irish North Western Railway.

The route north of Omagh gained importance with the opening of the line from Portadown in 1861.

4 – 4 – 0 No 170 "Errigal" ready to leave Londonderry Foyle Road with an express for Belfast in September 1957.   (J C W Halliday)

It was only after the amalgamation, which formed the Great Northern, that Belfast – Portadown – Omagh – Londonderry began to be looked upon as a through main line. In 1907, the final 7½ miles from St Johnston into Londonderry were widened to double track but were reduced again to single in 1933. This apart, the whole of the Irish North Western was always single track.

In 1910, weekday departures from Londonderry Foyle Road were as follows:

7.05am	FO	Strabane (slow)
7.30am		Dundalk (slow)
9.10am		Strabane (slow)
11.10am		Dundalk (slow)
12.15pm		Belfast (slow)
2.35pm		Dundalk (slow)
3.30pm		Belfast (fast)
4.20pm		Clones (slow)
5.00pm		Belfast (semi fast)
9.30pm	A	Portadown (semi fast)

FO = Fridays only + Letterkenny Fair Days, connecting with the 7.53am Strabane to Letterkenny.

A = Through sleeping cars to Dublin Amiens Street (arrive 5.10am). This train runs seven days a week and is the only Sunday departure from Londonderry Foyle Road.

After 1921, the Border altered established flows of traffic and Belfast – Portadown – Omagh – Londonderry became the main line with Dundalk – Clones – Enniskillen – Omagh as more of a long branch. For most purposes, the "Derry Road" was an internal Northern Ireland route but it crossed the Foyle north of Strabane and ran through County Donegal for nine miles before re entering Northern Ireland. This anomaly helped to keep Customs officers in employment. During the Second World War, it prevented military traffic from using this stretch of line. Materials landed at Londonderry, bound say for Enniskillen, had to be winched across the Craigavon Bridge to Waterside station and then worked via Cookstown and Omagh. Military personnel had to travel in "civvies" or wear something over their uniform and/or ride in the guard's van. Only an enthusiast would think of changing at Strabane onto the sparse service over the narrow gauge. There was talk of regauging this line to bypass the problem but nothing ever came of it.

When the Great Northern was dissolved in 1958, the nine miles of track and three stations became the property of CIE, who employed the signalling, station and permanent way staff. All the trains were operated by the UTA until the NI Government closed the line in 1965.

A BUT formation on a Belfast service at Londonderry Foyle Road in August 1959. The leading car is one of the 901 to 908 series. *(Peter Sunderland)*

Carrigans Station looking north in 1959. The track had been singled in 1932. The concrete building dated from about that time and included a Customs hall. After 1958, Carrigans was a CIE station served only by UTA trains. *(Geoffrey Lewthwaite)*

4 – 4 – 0 No 132 entering Londonderry Foyle Road with a train from Belfast in July 1957. The goods depot is on the site of the Londonderry & Enniskillen terminus of 1847 – 1850. *(Peter Sunderland)*

A BUT railcar set crossing the River Mourne, just south of Strabane, on a Londonderry to Belfast working in August 1959. The narrow gauge bridge, for Stranorlar, is just beyond the GNR structure. *(Peter Sunderland)*

No 68 exchanging tokens as it heads a Londonderry to Belfast train through Victoria Bridge in August 1959.

(Peter Sunderland)

4 – 4 – 0 No 205 "Down" (built 1948 to a 1915 design) outside Omagh Shed in August 1956. The shed was about ¼ mile east of the station. The Belfast line is to the left.   *(Peter Sunderland)*

4 – 4 – 0 No 202 "Louth" pulling away from Enniskillen with the 11am Bundoran in August 1956. The goods yard was the Londonderry & Enniskillen terminus, before the line extended southward.

(Peter Sunderland)

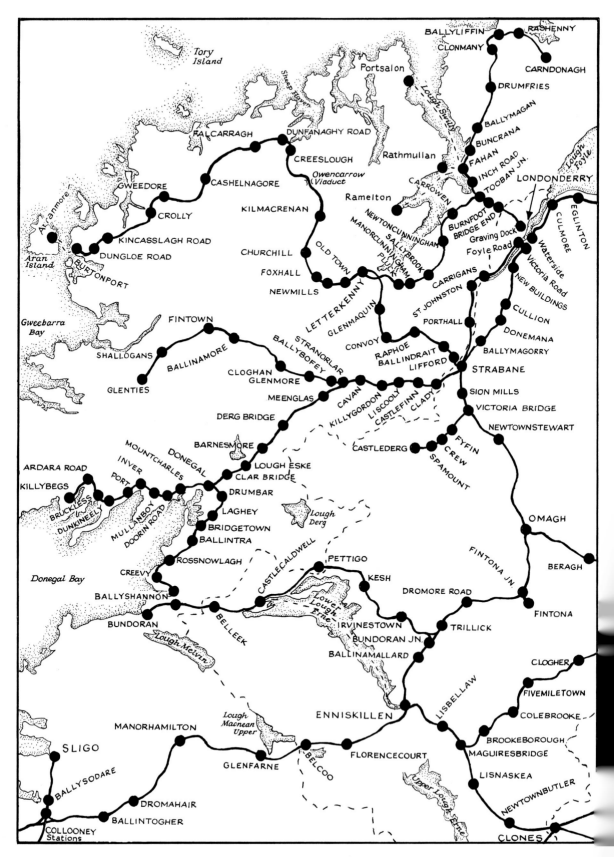

Ballybay Station looking west in 1958. The passing loop could not pass two passenger trains. There was a bay platform at the far end for the Cootehill train.
(J C W Halliday)

Railcar C2 at Monaghan Road on 2 August 1958, returning from Cavan to Dundalk.
(J C W Halliday)

North of Keady, the 11 arch Tassagh Viaduct stands as a memorial to the Armagh – Castleblaney line.

(Alan Young)

QG class 0 – 6 – 0 No 154 with a couple of six wheelers at Cootehill in 1939. *(W A Camwell)*

Armagh to Castleblaney

The last section of the Great Northern to open was also the first to close. The 18¼ mile branch from Armagh to Castleblaney was born out of inter railway rivalry and killed off by the Border.

Since 1875, the Midland Great Western Railway had worked as far north as Kingscourt. Any extension of the Kingscourt branch would amount to an invasion of Great Northern territory. In 1893, the MGWR deposited a Bill for such an extension to Armagh, Dungannon and Cookstown. This was almost a revival of the 1838 Royal Commission proposal for an "inland" main line between Dublin and Belfast. If the scheme were to go ahead, all the way to Cookstown, then it would become possible to get from Dublin to Belfast or Londonderry without touching the Great Northern. In the face of GN opposition, the MGWR decided to back down. To avoid an expensive Parliamentary battle, it agreed not to seek any extension of its own line north of Kingscourt. In return, the GN promised no incursion into MGWR territory.

In 1900, the MGWR broke the truce by offering substantial backing to the nominally independent Kingscourt, Keady & Armagh Railway, which had gained Parliamentary approval despite GN opposition. The response of the GN was to throw more money at the Keady project than the MGWR

was offering. By an Act of 1902, the southern terminus of the proposed new line was changed from Kingscourt to Castleblaney. The name of the undertaking was similarly amended.

The line proved difficult to build. It included three stone viaducts at Milford, Tassagh and Keady. The summit at 613ft above sea level was the highest point reached on the Irish standard gauge. After five years, the Company sacked the contractor, took over his plant and completed the line with direct labour. The eight miles from Armagh to Keady opened on 31 May 1909, with five trains a day, weekdays only, mostly timed to connect to and from Belfast. The remaining 10¼ miles followed on 10 November 1910. The following year, the Castleblaney, Keady & Armagh Railway was absorbed into the Great Northern. In 1913, two of the former contractor's locomotives, 0 – 4 – 0ST "Kells" and 0 – 6 – 0T "Mullingar" were taken into GN stock. They survived until 1930.

The Railway they had helped to build, however, lasted as a through route only until 1923 when the Keady to Castleblaney section was closed to all traffic. This move avoided the need for Customs posts at Carnagh and Creaghanroe. Keady itself retained a passenger service to Armagh until 1931. Goods traffic continued until the "Great Closure" of 1957.

The Cootehill Branch

The 7½ mile branch from Shantonagh Junction was part of a greater ambition by the Dundalk & Enniskillen to reach Cavan. In the event, Cavan was approached via Clones and the D&E branch went ahead only as far as Cootehill. It opened on 18 October 1860.

Branch trains left from a bay platform at Ballybay, 2 miles east of Shantona Junction. For most of its life, the branch had four passenger trains each way, some of which ran "mixed". Traffic was worked from Cootehill, where there was a small engine shed. There was no Sunday service.

Fair Day at Cootehill was the third Friday of the month. Sufficient wagons had to be available for the early afternoon "mixed" to dispatch livestock, some of which would be heading for the cross channel steamers. Carriages were needed for drovers with, possibly superior, accommodation for dealers. These resources had to be moved around the network to serve Fair Days at the various towns.

The line lost its passenger service in the fuel crisis of March 1947. Regular goods traffic ceased at the same time but livestock was carried on the monthly fair day until 1955.

The Branch to Carrickmacross

The 6¾ mile line was authorised in 1881 and opened on 31 July 1886. It left the Irish North Western at Inniskeen, where a third platform was provided for branch traffic. The maximum gradient was 1 in 100.

Carrickmacross boasted a single platform terminus with overall roof, a goods yard, engine shed and wooden signal box. In 1910, trains left Carrickmacross at 8.05 and 10.15am, 3.10 and 6.15pm, returning from Inniskeen at 9.33, 1.20, 4.40 and 7.40. All trains had good connections to and from Dundalk. The second Thursday of the month was Fair Day in Carrickmacross but, on all Thursdays, the train made an additional trip at 8.35am from Inniskeen coming straight back at 9.00 from Carrickmacross.

The branch closed to passengers in March 1947 but remained in use for goods until the last day of 1959, when CIE rid itself of the final remnants of the Irish North Western. There were occasional passenger movements after 1947, including for sporting events. Through carriages were sometimes run off the branch to Pettigoe, for the Lough Derg pilgrims, and Bundoran. The last passenger train was a DMU on 19 December 1959, chartered by the Irish Railway Record Society.

2G No 153, built 1903, running round its train at Carrickmacross in June 1939.   *(W A Camwell)*

The Fintona Branch

Fintona was the most important source of traffic between Omagh and Enniskillen. But extension southwards had left the town at the end of a ½ mile branch line. Freight was worked by steam but, from 1854 until 1957, the passenger service relied on a horse.

The first Fintona carriage looked a bit like a stagecoach with three compartments, one first class and two second. Third class passengers were accommodated on the roof, there being no over bridges on the branch. With six a side seating, total capacity was 72 passengers.

In 1883, this vehicle gave way to something like a conventional tramcar. This had 24 seats on the lower deck, which was divided into first and second class saloons. Third class comprised 24 longitudinal wooden seats on the open top deck. There was a baggage trailer for use when parcels or luggage traffic was heavy.

By the 1950s, the horse tram was something of

an institution making 11 return trips each weekday between Fintona Town and Fintona Junction in connection with trains on the Omagh to Enniskillen line. The tram was driven by station staff, based at Fintona Town. The horse, whether male or female, was always called "Dick". The last "Dick" was a male, born in 1942, who worked the tram from 1945 until closure. If he was out of traffic, a local carthorse was hired. There was no relief for the tram, which was damaged in an accident on 17 January 1953. Whilst it was away for repairs, Dick hauled a goods van but passengers had to walk along the track. Normal service resumed on 2 April.

The branch fell victim to the "Great Closure" of 30 September 1957. It had already been decided that the tram would be given to Belfast Transport Museum. It is now at Cultra. "Dick" was advertised for sale. The matter attracted some publicity and donations were received by the Ulster Society for the Prevention of Cruelty to Animals. They bought him and found him a retirement home.

"Dick" is ready to leave the Junction for Fintona Town on 7 September 1957.
(J C W Halliday)

Fintona Town on 7 September 1957 with the tram under the overall roof. Goods traffic was worked down the branch by steam locomotives.
(J C W Halliday)

"Dick" is harnessed up for the return to Fintona Town. Two young ladies take a last look at him whilst waiting for their connecting train.

(Peter Sunderland)

Fintona Junction Motive Power Depot. After bringing
in its branch train, the horse usually took refuge inside
the shed so as not to be frightened by the connecting
steam train. *(H C Casserley)*

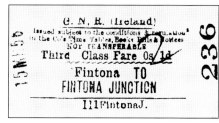

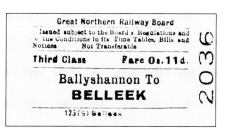

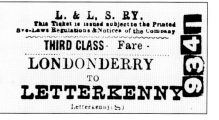

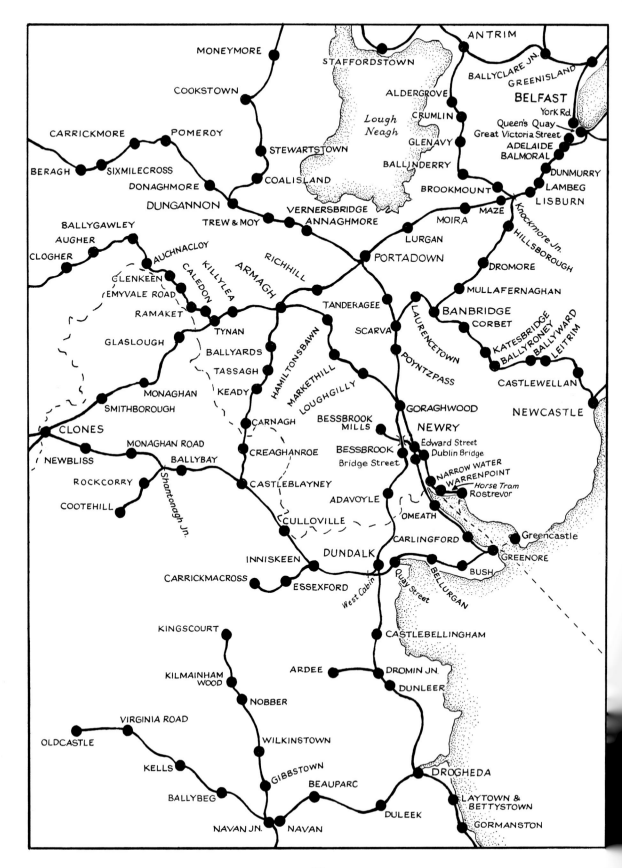

The Portadown, Dungannon & Omagh Railway

170 "Errigal", on Londonderry to Belfast express, waits for a booked crossing at Pomeroy on 7 September 1957. *(J C W Halliday)*

This enterprise completed what became the "Derry Road", the Great Northern route from Belfast to Londonderry. The first 13¾ miles from Portadown to Dungannon were authorised by an Act of 1847 but capital was not forthcoming and the powers had to be renewed in 1853. Work finally got underway in 1855, thanks to financial support from the Ulster Railway. The contract was awarded to the English firm of Fox, Henderson & Co but they became insolvent after a year and William Dargan had to be asked to complete the project.

The line opened on 5 April 1858 but, for the first few months, passenger trains were not allowed to use the junction at Portadown because it wasn't properly signalled. A temporary platform was erected on the Dungannon side of the junction, whence customers could walk to the Ulster Railway station.

The facilities at Dungannon were also of a temporary nature, the line terminating about a mile short of the town pending the extension towards Omagh. This was authorised in August 1858 and opened on 2 September 1861. It joined the pre – existing Londonderry & Enniskillen Railway at the entrance to Omagh Station.

The Portadown, Dungannon & Omagh was worked by the Ulster Railway. As soon as it was completed through to Omagh, it was leased by the Ulster. The local Company received its rent until 1876 when it was taken over completely, just three months before the Ulster Railway itself became part of the Great Northern.

The Portadown to Omagh route had a maximum gradient of 1 in 71. It reached a summit at mile post 26½, midway between Pomeroy and Carrickmore. The biggest engineering work was the 814 yard Dungannon Tunnel, through which the line approached the town of Dungannon from the south east. The tunnel wasn't required to surmount any physical obstacle but merely to appease the sensibilities of Lord Northland, who exercised sufficient influence that he could have threatened success of the Bill, had his objections not been upheld. The line between Portadown and Omagh was single throughout until 1902 when double track was completed from Portadown to Trew & Moy. Dungannon to Donaghmore followed in 1906. This latter stretch reverted to single in 1936, the former in 1959.

At Omagh, the physical junction between the Belfast and Enniskillen lines was at the South box. No 192 "Slievenamon" is taking water at the up platform which continued a short way along the Belfast line. In the lower photo it is leaving for Belfast. It is not clear what No 190 "Lugnaquilla" is doing on the down line, but they did exchange carriages with Enniskillen trains. *(Peter Sunderland)*

"Commendable if perhaps too late" is how *Trains Illustrated* described the revised "Derry Road" timetable introduced on 1 October 1957. This was the morning after the "Great Closure", which had released some of the AEC railcar diagrams. In addition, the BUT cars were entering service allowing the entire Belfast to Londonderry passenger timetable to be dieselised.

In the old timetable, the first train from Londonderry had been the 6.55am with portions for Dundalk (arrive 12.05pm) and Belfast (arrive 11.00am). The new timetable offered two diesel trains: The 6.50am ran all stations to Strabane, then Omagh only. The 7.15am was first stop Strabane, then all stations to Dungannon, then Portadown and Belfast (arrive 9.55). 2 hours 40 minutes instead of the previous 4 hours 5 minutes. In place of the 9.30am "Restaurant Car Express", a diesel left Londonderry at 10.00, made exactly the same eight stops and arrived Great Victoria Street at 12.50, five minutes ahead of its steam predecessor. In similar vane, the 12.15pm steam "Express" was replaced by a 1.05pm diesel, which had travellers into Belfast at 4pm instead of 3.55. Altogether, there were five through trains each way, two on Sundays, between Belfast and Londonderry plus a handful of shorter workings.

The principal goods movements were overnight with trains booked to leave Belfast Mondays to Fridays at 8.20pm for Londonderry, 8.50 for Omagh, 10.00 for Londonderry and 12.45am for Cookstown.

There was also a train starting from Portadown at 3.05am, which was presumably a continuation of the 8.10pm from Dublin, due into Portadown at 2am.

The Cookstown Branch

In 1874, the Ulster and the Portadown, Dungannon & Omagh Companies jointly promoted a 14½ mile branch from Dungannon to the market town of Cookstown. By the time of opening on 28 July 1879, the project was part of the Great Northern. The track was single and the steepest gradient 1 in 70.

The station at Cookstown was alongside that of the Belfast & Northern Counties Railway, which had opened in 1856. The BNCR had not objected to the Dungannon & Cookstown but had obtained safeguards, one of which was that income from Belfast to Cookstown traffic should be pooled. The two routes were of roughly equal distance.

The passenger service ended on 16 January 1956, but goods traffic continued until the end of September 1959, when the branch was closed completely beyond Coalisland. A 5 mph speed limit was then imposed on the remaining section, which was worked as a siding from Dungannon Station. Dungannon Junction was abolished, the double track from Dungannon Station becoming one track for Omagh and one for Coalisland. The truncated branch survived in this form until the first weekend of January 1965, when the Ulster Transport Authority abandoned rail freight throughout Northern Ireland.

Passing Dungannon Junction in August 1958, the Cookstown branch to the right. *(Peter Sunderland)*

Cookstown with the GN terminus on the left and the NCC immediately to the right thereof. The main buildings survive, respectively as a hockey club and a chinese restaurant.

On his first encounter with the GN in 1955, our contributor noted the low volume of goods traffic – at least compared to what he was used to at Keighley. But he managed to find some. 0 – 6 – 0 No 7 shunting at Omagh in 1956
(Peter Sunderland)

By August 1959, No 200 "Lough Melvin" had become 65 of the UTA. It is ready to leave Strabane with the afternoon goods to Londonderry.
(Peter Sunderland)

QLG class 0 – 6 – 0 No 109 shunting Dungannon goods yard in August 1959. The lines to the left of the loco are the double track to Dungannon Junction. But, about this time, Dungannon Junction was abolished, after which the left hand track was for Omagh and the right hand for Coalisland
(Peter Sunderland)

Coalisland Station after closure to Cookstown in 1959. The signalling has been removed. Only the nearside track is in use as this forms a loop with the goods yard, which is off the right of the photo. The truncated branch lasted until the beginning of 1965, thanks to traffic in sand and bricks.
(Desmond Coakham)

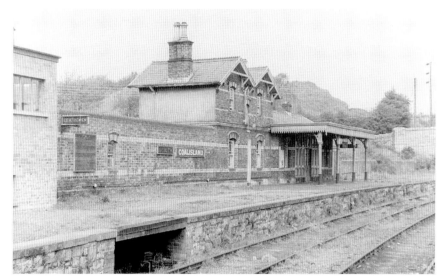

C3, with luggage trailer, calls at Beragh with an Omagh to Portadown local in July 1957.
(Peter Sunderland)

Omagh had two goods yards, one at the station and the other at the end of the short Market Branch. This began at Market Branch Junction. Morning and afternoon 'trips' were worked between the two, the loco propelling from Market Branch Junction.

Difficult Trading Conditions

4–4–0 No 73 shunting at Pettigo with the Bundoran to Enniskillen goods in 1955. In the 1949 working timetable, this train left Bundoran at 1.15pm, spent 2 hours 35 minutes at Ballyshannon, whence it ran "mixed" to Pettigo. It then meandered on as a goods, finally reaching Enniskillen at 8pm.

(Peter Sunderland)

The First World War gave way to the "Troubles", which lasted until 1923. First the Republicans fought the British. Then, after the Treaty of December 1921, a faction fought against the new Irish Free State. The Railways were often in the front line, suffering much damage and interruption.

As a result of the Treaty, the Great Northern became an international concern. About two thirds of its mileage lay in Northern Ireland, the rest in the Free State. The Border was crossed by railways in 20 places, 15 of them on the Great Northern. Mercifully, six of these came in quick succession between Clones and Cavan with no stations on the brief incursions into Northern Ireland. There has never been any restriction on cross border travel. Even so, the Border did interrupt established markets and patterns of travel.

Very soon, the Irish Government imposed different Customs duties from those of the United Kingdom and arrangements had to be made to enforce these along the Border and at sea ports. Facilities were provided for Customs Officers on the stations at Dundalk and Goraghwood, Glasslough and Tynan, Clones and Newtownbutler, Ballyshannon and Belleek, Pettigo and Kesh, Porthall and Strabane, Carrigans and Londonderry Foyle Road. In some cases, trains crossed the border two, three or even four times in the course of a

simple journey. The Customs examinations could be lengthy, even though they were only concerned with Revenue. It has never been necessary to carry a passport. The controls were abolished with the advent of the single European market in 1993 but that came far too late for all but one of the railway lines.

On top of these essentially local problems, the Great Northern had to face the more general hazard of road competition. During the 1920s, unregulated bus and road haulage firms grew rapidly, threatening the established business of the Railways. In some areas this didn't really matter as there was enough traffic to sustain both modes. But, in many parts of Ireland, there was scarcely enough business for one form of transport.

Left to market forces, the Railways would suffer very badly as they had the highest fixed costs. They also paid better wages, observed higher safety standards and accepted responsibility as common carriers and as part of a national, even international, network. The new competitors could just pick off business which suited them. The Railways, by contrast, were expected always to be there when needed but to manage without state aid.

Governments realised that things couldn't continue unregulated and, around 1930, legislation was passed in both parts of Ireland bringing road

H M Customs & Excise, Customs Station and Boundary Post, Goraghwood. AEC railcar 615 restarts for Dublin in July 1957. *(Peter Sunderland)*

The first Great Northern bus ran between Drogheda Station and Town Centre on 28 January 1929. The network expanded quickly but the Northern Ireland routes had to be given up to the Road Transport Board in 1935. No 257 was built in 1938 at Dundalk Works, one of a long line of GNR buses with Gardner engines. It ran until 1955.

(J C W Halliday)

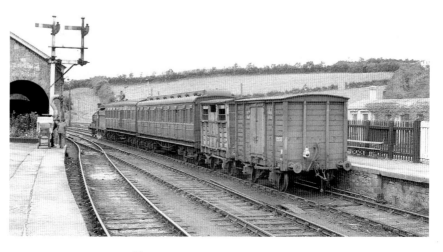

A "mixed" formation at the west end of Clones Station in September 1957. The cattle wagon is a reminder how important this class of traffic was, when meat was transported "on the hoof" both within Ireland and by the cross channel ships.

(J C W Halliday)

transport under a measure of control and allowing the Railway Companies to enter the road transport industry. Very soon the Great Northern had a fleet of 91 buses and 33 lorries, many of them bought from private operators along with the goodwill of their operations.

The policy of the Irish Free State was to allow the Railways this monopoly over all surface transport. In 1935, Northern Ireland embarked on a rather different course, nationalising virtually all road transport. The Railways were compelled to sell their road fleets to the Northern Ireland Road Transport Board in return for a promise that the latter would co-ordinate its operations with the Railways. In practice, it competed for traffic with the Railways who felt very aggrieved. Having bought out the private competition and introduced a measure of co-ordination, they faced renewed competition from a state monopoly.

Before the Government could carry out a promised review, the Second World War had broken out and road transport was severely restricted. Northern Ireland was at war and supplied with fuel. The Irish Free State was neutral and subject to the severest shortages. Railways with no cross border links could operate only the most restricted service. But the Great Northern had fuel and was able to profit from the additional wartime traffic. It was not under Government control so it was able to retain the extra revenue.

After the War, the Company soon found costs rising faster than income. In January 1951, it conceded that its resources were exhausted and began to serve notice on staff. The threatened shut down was avoided when the Governments of Northern Ireland and the Republic agreed to buy the Railway jointly. It took until 1 September 1953 to set up the Great Northern Railway Board. Despite the difficulties, the Great Northern had introduced the non stop "Enterprise Express" in 1947. It had built 15 new steam locomotives in 1948 and 20 diesel railcars in 1950/51.

The GNR Board persuaded the two Governments to authorise further railcars but failed to get a decision on diesel locomotives. Instead, during 1955, the Northern Ireland Government announced its intention to close 115 route miles representing its share of three cross border routes; Portadown to Clones, Omagh to Clones and Bundoran Junction to Bundoran. The proposal was made against the advice of the GNR Board and against opposition from the Irish Government, which would be left with a number of practically useless stretches on its side of the Border.

The Northern Ireland Transport Tribunal held a public enquiry at which every party represented was opposed to closure. Yet the Tribunal found in favour of the Government, which had not sent a representative to give evidence. In June 1957, the Government instructed the Board to cease operating the threatened lines after 30 September.

On 22 July, the GNR Board staged a last act of defiance by putting one of the new BUT railcars on a Belfast to Enniskillen roster, reducing the journey

AEC railcar No 619 at Clones on 7 September 1957 with an adapted coach as centre trailer and, most probably, 618 at the far end.   *(J C W Halliday)*

204 "Antrim" arriving at Clones with the westbound "Bundoran Express" in July 1957.

(Peter Sunderland)

209 "Foyle" on arrival at Dublin Amiens Street with the "Enterprise Express" in May 1956.

(Peter Sunderland)

time from 3½ to 2½ hours each way. The train only ran for ten weeks, pending closure between Omagh and Enniskillen.

Faced with the loss of all through traffic, the Irish Government hurried legislation to permit closure to passengers of a corresponding 84 route miles in the Republic. This could not be achieved in time so, for the first 13 days of October, passenger trains ran over certain "stump lines".

The "Great Closure" of 1957 also took the Sligo, Leitrim & Northern Counties Railway. It created the largest area in the British Isles without railways. Over the next few years, the rail – less area grew larger with the loss of the Cavan & Leitrim, the County Donegal and the remaining GN goods lines south of the Border. Perhaps the biggest loss still to come was the GN "Derry Road" from Portadown to Londonderry.

Already, during Summer 1957, the Northern Ireland Government had said that the "Derry Road" would have no long term future. Despite this, the GN Board deployed some of the new BUT cars to offer an accelerated timetable between Belfast and Londonderry from 1 October 1957.

The days of the Great Northern Railway itself were already numbered as the Northern Ireland Government had given notice to terminate the Joint Board on 30 September 1958. What was left of the Railway was divided between CIE and the UTA. Each class of motive power and rolling stock was divided in more or less equal numbers between the two organisations.

The Benson Report

The Northern Ireland equivalent of the Beeching Report was commissioned under the chairmanship of Sir Henry Benson, a London Chartered Accountant. It was delivered in Summer 1963, accepted by the Government and implemented in full so far as the former Great Northern Railway was concerned.

On 30 October 1964, the Transport Tribunal approved closure of the line between Portadown and Londonderry, the freight only Coalisland branch, the Warrenpoint branch and the four stations on the Main Line south of Portadown. Legal action by Tyrone County Council earned a six week reprieve for the "Derry Road" but the other closures took place on the first weekend of 1965.

On Sunday 14 February, the 6.30pm from Londonderry and the 8.05 from Belfast were the last passenger trains over the 75 mile "Derry Road". As usual, both were worked by BUT railcar formations whose subsequent redeployment virtually ended the use of steam on passenger trains in Northern Ireland.

The Great Northern Antrim branch had closed to passengers in 1960 but had remained open for goods. From February 1965, it was kept in use by a CIE train carrying goods from Dublin to County Donegal. Hitherto, this traffic had gone to Strabane for onward movement by road vehicles belonging to the County Donegal Joint Committee. But from 1965, the nearest railhead to Donegal became Londonderry Waterside.

After 1965, all that remained of the Great Northern was the Dublin to Belfast Main Line, the suburban branch to Howth and freight only branches to Navan, Ardee (since closed) and Antrim (also now closed).

The subsequent revival of the Main Line will feature in *Part Two* in a chapter on the "Enterprise Express".

Railcar C2 pauses at Castleblaney on 2 August 1958, working the daily parcels and mail trip from Dundalk to Cavan and return. *(J C W Halliday*

The Bundoran Branch

PP class 4 – 4 – 0 No 42 is ready to leave Bundoran in Summer 1957, with the 10.30 to Bundoran Junction and Enniskillen in July 1957.
(Peter Sunderland)

Bundoran is a seaside town on the Atlantic Coast of County Donegal. Powers for a 36 mile branch were granted, by an Act of July 1861, to the independent Enniskillen & Bundoran Railway. The Irish North Western was authorised to invest in the project, also to work the line once it was built.

An extension to Sligo was authorised in 1862, the Company changing its name to the Enniskillen, Bundoran & Sligo Railway. The extension was never built. Bundoran opened on 13 June 1866 and was always a terminus.

The Bundoran Company maintained its separate identity until 1897. The line had always been worked by the Irish North Western and its successor the Great Northern, who finally absorbed it under an Act of 1896. Three years later, the G N consolidated its position in the town by taking over the Bundoran Hotel which it renamed the "Great Northern".

The branch began at Bundoran Junction, where a station had existed since the opening of the Londonderry & Enniskillen in 1854. In the intervening 12 years, this had been known successively as Lowtherstown Road, Irvinestown and Irvinestown Road. As Bundoran Junction it survived 91 years. It had a triangular layout but with platforms only on the two sides leading towards Enniskillen. There were signal boxes at all three

corners of the layout.

The Bundoran line was single with passing loops at Irvinestown, Pettigo and Ballyshannon. The steepest gradient was 1 in 100 but there were many sharp curves necessitating speed restrictions. The route was scenic, running close to the north shore of Lough Erne for 17 miles.

Three trains a day made up the year round passenger service, weekdays only. In 1910, these left Bundoran at 7.15, 11.35am and 5.25pm. Return from Bundoran Junction was at 9.10am, 1.12 and 7.05pm. There were connections to and from both Enniskillen and Londonderry. In two of the three instances, the north and south bound main line trains passed at Bundoran Junction, which thus had three trains in at once. The morning pair of main line trains passed further north at Dromore Road.

In mid summer, there was a through train in the late morning from Enniskillen to Bundoran. Between 1896 and 1914, tourists had the option of leaving Enniskillen earlier, at 10am, on the paddle steamer "Lady of the Lake" and joining the train for Bundoran at Castlecaldwell, where pier and station were adjacent. A corresponding connection was available off the afternoon train from Bundoran. Through rail and steamer tickets were available but the "Lady of the Lake" was withdrawn after the 1914 season. Lough Erne sailings were revived briefly in the early 1930s using the motor launch "Enniskillen".

Restaurant cars were introduced on the Irish North Western in 1927. In summer, the car went to Bundoran. It was attached to the 10.12 Dundalk to Londonderry that stopped at almost every station. This train included through coaches from Dublin to Bundoran. At Clones, the Bundoran portion was detached and joined by a through coach from Belfast. It then proceeded in advance of the Londonderry train stopping only at Enniskillen, Pettigo and stations thence to Bundoran. It was usually worked by 4 – 4 – 0 No 197 from Clones Shed which was more powerful than the Bundoran engines. On the return, the "Restaurant Car Express" left Bundoran at 3.15pm and was combined at Bundoran Junction with the 2.35 from Londonderry.

In the mid 1930s, three locomotives were shedded at Bundoran. Nos 52 and 53 for the passenger trains were 4 – 4 – 0s built by Beyer Peacock in 1892. No 32 for the goods was a 0 – 6 – 0 of 1894 vintage. Diesel railcar "C" was employed on the branch from 1934, running up to 1,000 miles per week including journeys to Enniskillen.

Between Clones and Ballyshannon, the Railway passed mostly through Northern Ireland but Pettigo Station lay just inside County Donegal. This meant that the Border was crossed four times. Just to make things complicated, during the Second World War, the Border was also a frontier between time zones with the United Kingdom observing "double summer time". From 1943, the Great Northern avoided Customs delays by having the newly named "Bundoran Express" run non stop through Northern Ireland. The 1955 summer timetable shows a complete train leaving Dublin Amiens Street at 8.45am with stops only at Drogheda, Dundalk, Inniskeen, Castleblaney, Ballybay, Clones, Pettigo and Ballyshannon. Arrival in Bundoran was at 2.00pm, 5¼ hours for 160 miles. The return working left Bundoran at 12.25pm, crossing the incoming train at Pettigo. The stop here was important for

Class PP 4 - 4 –0 No 42 pulling into Pettigo with a slow to Bundoran in July 1957. The long canopy on the nearside platform offered shelter for returning Lough Derg pilgrims who, on occasions, could fill the station. *(Peter Sunderland)*

C D Ry J Co
DAY TRIP
Third Class
Lough Eske
To
BUNDORAN
Via Donegal & B'shannon

44 at Belleek in 1957, heading for Bundoran. The "reduce speed" board warns of the Ern Bridge. UTA cattle wagon occupy the road above o Belleek Fair Day.
(Peter Sunderland)

No 42 has arrived at Bundoran Junction with the train from Bundoran, due in at 12.02. It then shunts across to the main line to make way first for the 12.00 Enniskillen to Omagh, for which people are waiting, and then the non stop "Bundoran Express", due to pass through the branch platform at 12.39. No 42 will leave for Enniskillen at 12.44.

(Peter Sunderland)

No 204 "Antrim" calls at Ballyshannon with the "Bundoran Express" from Dublin in August 1956.

(Peter Sunderland)

pilgrims praying and fasting on an island in Lough Derg

The 1955 timetable shows a daily pick up goods leaving Bundoran Junction at 7.00am to which it finally returned at 4.52pm after a leisurely trip shunting all branch stations and subjecting itself to frequent Customs examination.

There was never a year round Sunday service on the branch but Bundoran did attract a lot of Sunday traffic in summer. In the 1930s, there were five timetabled trains into Bundoran, three from Clones and two from Londonderry via the north curve at Bundoran Junction. These all arrived in the morning or early afternoon and returned during the evening. In addition, there were advertised and chartered excursions, which could result in up to eight engines appearing on Bundoran shed before the evening exodus. Some of the carriages had to be worked back to Ballyshannon to find siding space.

Some excursions involved crews in very long days. For a Cookstown to Bundoran trip, a Portadown crew would have to book on about 5am in order to leave at 6am with empty carriages for Cookstown. There they arrived at 7.10 in time to run round and prepare for the 7.30 departure. They had to run round again at Dungannon and again at Omagh, whence they ran non-stop to Belleek, where British Customs made sure that they were not exporting a trainload of contraband. A stop at

Ballyshannon allowed Irish Customs to make sure that they paid duty on it, if they were. Finally, Bundoran was reached at 11.20. There, after a bit of shunting, the crew could book off until it was time to prepare for the 7.05pm return.

Pausing only at Ballyshannon and Belleek for Customs and Bundoran Junction for water, the train reached Omagh at 9pm. But it was 11.30 before the weary passengers were deposited at Cookstown and 12.45am before the even wearier crew reached Portadown.

An occasional all day excursion from Belfast was marketed as the "Hills of Donegal". This travelled from Belfast to Strabane, where passengers changed onto the narrow gauge for Ballyshannon. The GN train then worked empty to Bundoran where the loco went on shed. The Bundoran branch train performed an extra short working to Ballyshannon, to retrieve the excursion passengers who had been "route marched" through the town from the County Donegal Station with a railwayman at their head. After a three hour break at Bundoran they boarded the main train for the 3½ hour journey back to Belfast

The Bundoran branch fell victim to the "Great Closure" of 30 September 1957. The Hotel passed to CIE in 1958 but was later sold on. It still functions as the "Great Northern", part of the McEniff Hotel Group.

The Warrenpoint Branch

PP class 4 – 4 – 0 No 76 on a Goraghwood train at Newry Edward Street in July 1957.

(Peter Sunderland)

The 10 mile branch of the Great Northern Railway left the Dublin to Belfast main line at Goraghwood. It passed through the important town of Newry before running alongside Carlingford Lough to reach the seaside town and port of Warrenpoint. The line was built in two parts. At first they were physically separate.

The Newry, Warrenpoint & Rostrevor Railway

As authorised by an Act of July 1846, this would have been an 8½ mile double track line, extending as far as Rostrevor. Capital was insufficient to permit more than a single line ending at Warrenpoint. The contractor was William Dargan, who had to accept debenture stock for part of the money owed to him.

The line opened on 28 May 1849. Five 2 – 2 – 2 tender engines had been ordered from Bury, Curtis & Kennedy of Liverpool, but the Company could not afford them all. They paid a penalty to cancel two of them and took until 1854 to pay for the other three. The engines had to work tender first in one direction, because the Company fought shy of the price quoted for two turntables. After a year, the Company subcontracted working the line to Dargan, resuming direct control in 1855.

The Warrenpoint & Rostrevor Tramway

The Railway never reached Rostrevor, at least not by steam. On 1 August 1877, a 3 mile horse tramway was opened from outside Warrenpoint Station to a terminus surrounded by trees, opposite the Mourne Hotel at Rostrevor. The single track was laid mainly along the roadside. The gauge was 2ft 10in. There were three intermediate passing loops. A short extension to Rostrevor Quay opened in October 1877.

Six passenger cars were supplied by Starbuck of Birkenhead, four open toastracks and two saloons. They were similar in size and appearance to some of those still running on the Douglas Horse Tramway, opened a year earlier, in the Isle of Man. There was also a goods wagon. The line closed in February 1915, after a length of track had been damaged in a gale.

The Great Northern Railway never owned the tramway but it certainly regarded Rostrevor as one of its holiday destinations. In 1899, the Railway purchased the Mourne and Woodside Hotels at Rostrevor as well as the Beach Hotel at Warrenpoint. The Woodside was sold after a few years but the other two were upgraded and renamed "Great

Class U 4 – 4 – 0 No 198 "Lough Swilly" at Goraghwood with a Warrenpoint to Belfast train in July 1957. *(Peter Sunderland)*

AEC railcar No 615 on a Belfast to Dublin run, being subjected to Customs examination at Goraghwood in July 1957. *(Peter Sunderland)*

T2 class 4 – 4 – 2T No 11 entering Goraghwood on Belfast to Newry working i July 1957. It is passing the Grea Northern Railway ballas quarry. *(Peter Sunderland*

U class No 198 "Lough Swilly" entering Newry Edward Street with a Warrenpoint to Belfast local in August 1957. The GN had five signal boxes in Newry: North and South (at either end of Edward St Station), Monaghan St (gate box), King St Junction (for Greenore) and Dublin Bridge.

(Peter Sunderland)

2 class 4 – 4 – 2T No 116 at Newry Dublin Bridge, with a local from Warrenpoint in July 1957.

(Peter Sunderland)

Northern". The Warrenpoint establishment was sold in 1922 but the Rostrevor "Great Northern" remained in Railway ownership until sold by the Ulster Transport Authority in 1966.

The Newry & Armagh Railway

This began life as the Newry & Enniskillen, an ambitious 71 mile project, authorised in 1845. The route from Armagh to Enniskillen was eventually built but not by this Company, which struggled even to complete the 3¼ mile stretch from Newry to Goraghwood, opened on 1 March 1854.

By an Act of August 1857, powers were renewed for the Goraghwood to Armagh section, the Company's name being changed to the Newry & Armagh to reflect its reduced ambition. The 17½ mile extension involved some heavy engineering, including the longest tunnel in Ireland at Lissummon, 1,759 yards. The line opened to a temporary station, just outside Armagh on 25 August 1864. It took a further six months before trains could run into the Ulster Railway's station. The Ulster had objected to the 1857 Act, and was able to extract a high price from its neighbour for use of Armagh Station.

The Town of Newry Connecting Railway

The Warrenpoint and Armagh Railways did not meet in Newry but had separate termini ¾ mile apart. The isolation of the Warrenpoint line was a severe disadvantage, particularly if it was to develop any goods traffic. A connecting line was discussed, even before the opening to Goraghwood in 1853. The mile link was authorised by the Newry & Armagh Act of 1857. It involved a movable bridge over the Newry Canal and five level crossings. It opened on 2 September 1861, when Newry Dublin Bridge replaced the previous terminus of the Warrenpoint line. Dublin Bridge Station had an overall roof spanning the single platform and loop line, but the roof was demolished about 1912.

The Great Northern Takeover

The Newry & Armagh was taken over by the Great Northern Railway in June 1879. One of the last transactions of the N&A had been to order a 0 – 4 – 2 loco from Sharp Stewart which they couldn't afford to pay for.

The Newry, Warrenpoint & Rostrevor remained independent for another seven years. During 1885, it began negotiating sale to the Great Northern. The

4 – 4 – 2T No 116 running round at Newry Dublin Bridge, having terminated with a train from Warrenpoint in July 1957. The bridge used to slide, prior 1937 when the Newry Canal closed to navigation.   *(Peter Sunderland)*

Q class 4 – 4 – 0 No 131 entering Goraghwood in 1957 with cattle from Newry to Belfast. Built by Neilson Reid in 1901, No 131 was withdrawn by CIE in 1963. It finally re-entered traffic on 5 November 2017 hauling a special from Whitehead to Belfast.   *(Peter Sunderland)*

Open and saloon cars at Rostrevor terminus, just beyond the Great Northern Hotel.

The fireman surrenders the single line staff to the signalman at Warrenpoint.   *(Peter Griffin)*

price was about a third what it had cost to build. The terms were confirmed by an Act passed in June 1886.

Warrenpoint Station was rebuilt in 1890, but the promise of a more central passenger station in Newry never materialised. According to the local press, nearly all NW&RR staff were dismissed upon take over but the Great Northern "very generously" offered to try and find new employment for some of them. There were no workers rights in those days. That said, the Great Northern was probably keener to keep some of the staff than it was the locomotives.

Of the six Newry & Armagh engines taken over, four were gone by 1887 and all by 1895. The Sharp Stewart 0 – 4 – 2 did not pass to the GN but ended up on the Belfast & County Down Railway. Of the four NW&RR engines, only two were taken over by the GN in 1886. They lasted six and 12 years with the new owners.

In 1910, there were nine departures each weekday from Warrenpoint. Four of these continued to Armagh, after a wait at Goraghwood of up to 20 minutes. The first train of the day, at 7.45am was through to Belfast. The last at 8.45pm went only as far as Newry Edward Street. There were two trains starting at Newry plus the "Boat Train", through from Greenore to Belfast. There were three departures from Warrenpoint on a Sunday, but no service over the Goraghwood to Armagh section. In the 1954 timetable, the standard of service from Warrenpoint and Newry is similar.

The passenger service between Goraghwood and Armagh had ceased during the 1933 strike, when the line was closed completely beyond Markethill. The remaining section from

Goraghwood carried a modest amount of freight until the end of April 1955.

The abandoned track bed, from Goraghwood to Lisummon Tunnel, was surveyed in the late 1970s by Northern Ireland Railways, who were looking for a way of dumping asbestos contaminated DMU vehicles. Track would have been re-laid into the tunnel and the vehicles propelled inside. Then the tunnel mouths would have been sealed with concrete and the track back to Goraghwood removed again. In the event, the rolling stock was buried in a quarry near Crumlin instead.

Decline and Closure

Sunday trains between Newry Edward Street and Warrenpoint were withdrawn in March 1956. Then, from the start of the 1961 – 62 winter timetable, the weekday passenger service was reduced to just one train each way, run for the benefit of school children at 8.40 am from Warrenpoint and 3.50 pm back from Newry. There was still year round freight and a more comprehensive passenger service in summer, augmented by excursions. Newry retained a respectable service to Goraghwood and beyond.

The branch was condemned in the 1963 Benson Report, which was accepted by the Government who promised that a motorway would be built from Belfast to Newry. It never has been.

The last passenger train left Warrenpoint at 8.40am on Saturday 2 January 1965. It was hauled by 0 – 6 – 0 No 40 (GN No18) which also worked the last goods out of Warrenpoint later the same day. The last train from Newry Edward Street was a six car DMU to Belfast Great Victoria Street on the Sunday evening.

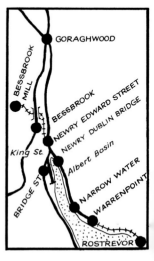

South from Goraghwood, the Dublin line rises at 1 in 113 whilst the Warrenpoint branch fell at 1 in 138. The 4 – 4 – 2T, seen at Goraghwood on the title page, is now photographed from the Dublin train, making its way towards Newry.
(J C W Halliday)

The Great Northern Hotel, Rostrevor, which was bought by the Railway in 1899. It was one of six former railway hotels in Northern Ireland, acquired by Hastings Group in 1971. They quickly fell victim to the "troubles" and the only one still open is the Slieve Donard at Newcastle.
(Martin Bairstow Collection)

Markethill Station in 1958. It had lost its passenger service in 1933, when the line was closed between here and Armagh. Goods traffic had then continued from Goraghwood until 1955. *(J C W Halliday)*

The GN obtained its ballast from its own quarry, at the north end of the Goraghwood Station. The hopper wagons were advanced by contemporary Irish standards.
(Peter Sunderland)

4 – 4 – 2T No 142 passing Narrow Water Castle with a local for Warrenpoint in July 1957.

(Peter Sunderland)

Warrenpoint on 12 September 1954 with 0 – 6 – 0 No 153, built 1903 and WT class 2 – 6 –4T No 7, built 1947. The WT was on loan from the UTA.　　　*(J C W Halliday)*

The Armagh Disaster 1889

On 12 June 1889, the Great Northern suffered a serious accident, which turned out to be a landmark in terms of legislation to enforce basic safety standards.

88 people, mostly children, were killed making Armagh the fourth worst railway accident in the British Isles after Quintinshill (1915), Harrow & Wealdstone (1952) and Lewisham (1957).

The single line from Armagh to Newry climbed at 1 in 75 to a summit, which was reached before the first station at Hamiltonsbawn. Traffic control was by staff and ticket, reinforced by a time interval. If a passenger train left Armagh on a ticket, another one could follow with the staff after an interval of 20 minutes.

On the fateful day, a heavily laden excursion for Warrenpoint left Armagh 15 minutes late at 10.15am. An ordinary passenger train was booked to follow at 10.35. The excursion comprised 2 – 4 – 0 No 86, a brake van, 13 six wheel carriages and another brake van. The locomotive stalled a short distance from the summit.

The guard should have walked back along the track to place detonators, which would serve as a stop signal to the following train. If the excursion remained unable to restart, the following train could have proceeded at caution until it reached it. Then, with any luck, the two engines would be able to get the combined train moving. Failing that, the whole thing would have to remain protected until someone could summon assistance via the nearest telegraph.

That is what the rules said but, in charge of the excursion and riding on the footplate was the chief clerk from Armagh Station and he thought he knew better. He instructed the train crew to uncouple between the fifth and sixth carriages. The idea was that the engine would take the front five to Hamiltonsbawn, then return for the remaining ten.

The train was fitted with a non automatic vacuum brake. This meant that, once they were no longer connected to the engine, the rear ten carriages were held only by the hand brake in the rear guard's compartment. It is not known whether this brake had been tampered with by children travelling in this part of the train. The guard himself was at the lineside, trying to augment the braking power by placing stones under some of the wheels.

As the locomotive tried to start forwards, it set back a little giving a nudge to the uncoupled rear portion, which began to roll down grade towards Armagh. The carriage doors had been locked before leaving Armagh so the occupants were not able to jump out.

The 10.35 from Armagh comprised a horse box and five coaches drawn by 0 – 4 – 2 No 9, a veteran of 1858. The driver was doing quite well, attacking the incline at 30 mph, when he saw what was coming towards him. He got his own speed down to 5 mph but the runaway was going much faster. Three carriages were completely destroyed in the collision, which beside killing 88 passengers, injured some 260 more.

Loco No 9 landed upside down on the slope of an embankment. The driver was thrown onto the coal plate of the tender which remained on the track, still coupled to the horse box. The coupling between this and the five coaches of the 10.35 was severed and the two parts began rolling back towards Armagh. No further accident ensued because the guard was able to stop the coaches with his hand brake whilst the driver had recovered sufficient to apply the hand brake on the tender.

The resulting furore moved the Government into swift legislation. Under the Regulation of Railways Act, 1889, automatic continuous brakes became mandatory for passenger trains. These would apply at once if a train became divided. The Act also made block working compulsory on passenger lines. Under this procedure, the 10.35 could not have been signalled to leave Armagh until confirmation had been received by telegraph that the previous train had cleared the block section to Markethill. If the crew of the excursion had admitted defeat, they would have had to get a message to Armagh confirming their position and the fact that their train was protected. Only then could Armagh have despatched an assisting engine under emergency rules, proceeding at caution until it reached the red flag and detonators protecting the stranded train.

Board of Trade Inspectors had long argued for continuous brakes and absolute block working, also the interlocking of points and signals. Some Companies had moved on these improvements but others claimed they couldn't afford the cost. The Armagh Disaster pushed the Government into compulsion. Nearly all companies complied fairly quickly. Apart from the human aspect, the financial consequences of the accident had put a new perspective on the cost of these basic operating safeguards.

Warrenpoint was the seaside railway terminus. The extension to Rostrevor was never built, except that between 1877 and 1915, the 3 mile gap was filled by a horse tramway. Two cars are seen outside the Great Northern Hotel in Rostrevor.

The Banbridge Lines

The small market town of Banbridge, County Down lies some five miles east of the Dublin to Belfast main line. By 1880, it was served by three branches of the Great Northern, all of which had been promoted independently.

The Banbridge Junction Railway

Incorporated by an Act of August 1853, the Banbridge, Newry, Dublin & Belfast Junction Railway had powers to build 6¾ miles of single track from the main line at Scarva. The route travelled northeast to Lawrencetown where it found the Valley of the Upper Bann. The contract was awarded to William Dargan. Work began in August 1854 but stopped in January 1856. A further Act of June 1856 increased the borrowing powers, at the same time shortening the name to the Banbridge Junction Railway. Work resumed in the summer of 1858 after an injection of capital from the Dublin & Belfast Junction Railway.

The line was inspected and approved by Capt Ross of the Board of Trade on 9 February 1859 but there was a delay agreeing terms for its operation by the Dublin & Belfast Junction Railway. Opening took place on 23 March 1859 and the branch was soon leased formally to the main line Company. It was fully absorbed into the Great Northern in 1877.

The Banbridge, Lisburn & Belfast Railway

The second approach to Banbridge was authorised in June 1858. The scheme was supported by the Ulster Railway, which was authorised to lease the line under a second Act in 1862. John Killen of Malahide was awarded the contract. Work started in June 1859 and was supposed to take two years but Killen lacked the capital to own sufficient plant. After a year, he was insolvent and the contract had to be re-let to Greene & King of Clones.

At first, the Ulster Railway had envisaged the branch operating only from Knockmore where there

would be an interchange station but with the physical junction only through a siding. This would avoid a facing point on the main line, which was something much feared in those days and avoided whenever possible. In the event, a conventional double junction was laid and, when the line opened on 13 July 1863, all trains ran through to Lisburn.

The Banbridge, Lisburn & Belfast Company received a rent from the Ulster Railway. There was continual disagreement between the two bodies, mainly over the state of the track. The local Company had insufficient means to do much about it. In 1873, a receiver was appointed to manage the BL&BR Company. Four years later, the line was sold to the Great Northern for a fraction of what it had cost to build.

The line was single with passing loops at Dromore and Hillsborough. The largest engineering feature was the seven arch stone viaduct over the River Lagan, just south of Dromore. The 514 ft summit of the line was reached north of Magherabeg. From there, the line fell as far as Newport, reaching a maximum gradient of 1 in 57, the steepest on the Great Northern Railway. At Banbridge, Lisburn trains used the same station as the Scarva line. This sensible arrangement, confirmed in the 1860 Act, replaced the original plan for a separate station with the tracks connected only through a siding.

The Banbridge Extension Railway

Banbridge became a through station with the opening of the 9 mile extension to Ballyroney in 1880. This had been authorised in 1861 as the Banbridge Extension Railway with support from the Ulster.

A contract to build the line had been signed with Moore Brothers but they were bankrupt even before work had begun. The replacement contractor was John Bagnall of Nenagh who began work in August

Railcar A, built at Dundalk in 1932, arriving at Banbridge from Scarva on 18 June 1953.
(H C Casseley)

1863, promising that the task would be finished in September 1864. That date was missed and, in October 1864, Bagnall stopped work until he received more money. The Company was having little success in raising funds. The Ulster Railway would not help. They even found some of their own shareholders transferring partly paid shares to people with no assets so as to avoid paying further calls.

The partially completed works stood derelict for 15 years. In 1876, the newly amalgamated Great Northern Railway surveyed the route and agreed to take it over. The purchase was ratified by the same Act of June 1877, which vested the other two Banbridge lines in the Great Northern. Work finally resumed during 1879 and the line to Ballyroney opened on 14 December 1880.

Ballyroney, in rural County Down, was not a great traffic centre. The main flows were coal into the district and agricultural produce out. Back in 1860, the promoters had seen Ballyroney only as a temporary terminus. Their real ambition had been Newcastle on the Coast. This had built up as a seaside resort following the arrival there of the Belfast & County Down Railway in 1869. In the late 1890s, the Great Northern proposed an extension from Ballyroney to Newcastle. The BCDR retaliated with a scheme of its own to bridge this same gap, at the same time seeking running powers to Scarva on the Main Line.

Compromise ensued and, under an Act of 1900, the Great Northern was given powers to build as far as Castlewellan with running powers for the last four miles into Newcastle. The BCDR was authorised to build from Newcastle to Castlewellan with running powers thence to Ballyroney. This latter concession was never taken up.

The route through to Newcastle was opened on 24 March 1906. It was single track with passing loops at Katesbridge, Ballyward and Castlewellan. The line climbed gently up the Bann Valley to a summit near Leitrim Station. It then descended mostly at 1 in 70 for six miles until it reached almost Sea level, two miles before Newcastle.

Train Services

In 1910, there were four trains each way between Belfast and Newcastle plus one only as far as Banbridge. On Sundays, there were two between Belfast and Banbridge. From Scarva, there were five trains on Weekdays with an extra on Mondays only. There was just one each way on a Sunday. There were additional trains between Castlewellan and Newcastle worked by the Belfast & County Down Railway. Besides the timetabled trains, there was excursion traffic to Newcastle from various parts of the Great Northern.

The Scarva to Banbridge passenger service was not resumed immediately after the 1933 strike, which lasted ten weeks from the end of January. It did, however, restart on 15 October 1934 using railbus "E". Later on, this gave way to railcar "A", which seated 48. For busy days, a lightweight trailer was provided with capacity for an additional 66.

As on other parts of the Great Northern, traffic increased dramatically during the Second World War but fell off afterwards. The Great Northern came under political pressure to close lines. Scarva – Banbridge – Castlewellan – Newcastle closed on 1 May 1955. A year later, Banbridge was completely without rail access when the route from Knockmore closed on 30 April 1956.

Leitrim Station, looking towards Newcastle in April 1953.   (H C Casserley)

Q class 4 – 4 – 0 No 121 has just left the main line at Knockmore Junction with the 6.20pm Belfast to Banbridge on 29 April 1956. It is passing Newforge Siding, which remained open until 1961 to serve a rendering factory.   *(R M Arnold)*

SG class 0 – 6 – 0 No 19 calls at Magherabeg with the 6.20pm Belfast to Banbridge on 10 August 1955.   *(R M Arnold)*

PP class 4 – 4 – 0 No 25 ready to leave Newcastle with the 9.30 to Belfast on 18 April 1953. The building to the right is the original BCDR station of 1869, superseded in 1906 by the more imposing structure with the clock tower.   *(R M Casserley)*

4 – 4 – 0 No 125 restarts the 5.22pm Belfast to Banbridge from Newport Halt on 6 August 1955. Around 1930 the GN opened a number of halts including three between Hillsborough and Banbridge at Ballygowen, Magherabeg and Ashfield. Newport, 1 mile north of Hillsborough, was a later addition in 1942.

(R M Arnold)

Hillsborough was an original station from 1863. 4 – 4 – 0 No 156 with the 4.55pm Belfast to Banbridge on 4 August 1955.

(R M Arnold)

4 – 4 – 0 No 74 pulling into Ashfield with the 2.45pm Banbridge to Belfast on 26 April 1955. The halt here had opened on 1 June 1930. *(R M Arnold)*

The Howth Branch

4 – 4 – 2T No 64 waiting for the "right away" at Howth on 6 August 1958.   (J C W Halliday)

Howth is a promontory overlooking Dublin Bay. It is connected to the mainland by a narrow flat strip of land. When the railway was promoted in the 1840s, there was hope of reviving the harbour and developing it as a cross channel port. This never happened. Instead, Howth grew as a seaside and commuter town.

Powers for the 3½ mile double track branch were given to the Dublin & Drogheda Railway in its 1845 Act. A single line opened on 30 July 1846 as far as a temporary station, ¾ mile short of Howth. The complete branch opened 10 months later on 30 May 1847.

The branch leaves the Belfast main line at Howth Junction, where there are diverging platforms on the two routes. Before 1912, the station here was just called Junction.

In 1859, there were seven trains a day to Howth, starting from Dublin Amiens Street. Fifty years later, there were 17. In 1906, the Great Northern had introduced a fleet of four steam railcars for the Dublin to Howth route. These could run alone or with a driving trailer. On Sundays, they provided a regular interval service, rare in those days, with departures from Howth every hour from 10am until 11pm.

The steam railcars were heavy on maintenance and not very comfortable. Carrying a steam engine and coal inside a passenger vehicle, they cannot

have been the cleanest mode of travel. They were withdrawn in 1913 and replaced by push – pull formations. These comprised 20 year old 4 – 4 – 0 tank engines sandwiched between a pair of carriages adapted as driving trailers. From his cab in the leading coach, the driver controlled the brakes and, by a system of rodding, operated the regulator on the locomotive. He used a telegraph instrument to relay the required position of the reversing lever, which was operated by the fireman. The unit could run with only one carriage, in which case the driver drove as normal from the footplate when the engine was leading.

Diesel railcars "C2" and "C3" ran on the branch in 1935 to 1937, coupled back to back. The branch was electrified on 23 July 1984 with class 8100 units running through to Bray.

The Hill of Howth Tramway
The Railway to Howth runs along the north side of the peninsular, virtually at sea level. To the south the land rises steeply to form the Hill of Howth. By the 1890s, people were gradually building houses up the hillside.

In 1897, the Great Northern obtained powers for a line, 5¼ miles in length, forming a loop between Sutton and Howth Stations, passing over the Hill of Howth. Gradients were severe, starting with half a

S class 4 – 4 – 0 No 173 "Galtee More" prepares to leave Howth for Dublin in August 1958.

(J C W Halliday)

No 2 at Sutton in summer 1958.

(Peter Sunderland)

mile at 1 in 16½ out of Sutton. There were long stretches at 1 in 20 as the line climbed from 25ft to 407ft above sea level and then dropped down again.

It was considered impractical to work the line as a conventional steam railway so it was built as an electric tramway. The gauge was 5ft 3in but there was no through working with the main line. The track was single with 11 passing loops. Each section was protected by a colour light signal. This gave a white light when the line ahead was clear but no light when a tram was approaching in the opposite direction. Consecutive movements in the same direction were permitted without need for the previous tram to have cleared the section.

The line opened from Sutton to the Summit on 17 June 1901 and down to Howth on 1 August. The service was frequent. Rolling stock comprised ten double deck tramcars. Nos 1 to 8 seated 30 on the lower deck and 37 on the open top. Nos 9 and 10,

built a year after the opening were slightly larger. There was only one class of accommodation. The depot was at Sutton. The 550 volt power supply was generated by the Company's own plant at Sutton until the 1930s, after which it was drawn from the public supply.

The line helped open up the Hill, allowing further residential development. It fed traffic onto the Howth Branch and catered for day trippers and sightseers. It survived dissolution of the Great Northern but was closed by CIE, with only 2½ weeks, notice on 31 May 1959.

Four of the trams have been preserved. No 2 is in California. No 4 is in the Ulster Folk and Transport Museum at Cultra and No 10 in the National Transport Museum at Crich, Derbyshire. No 9 stayed in the Dublin area but was allowed to rot until 1974, when restoration began. It is now displayed in the Irish Transport Museum at Howth Castle.

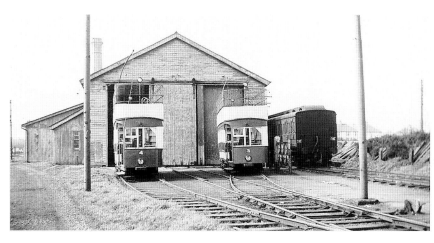

The tram shed was alongside the railway at the Dublin end of Sutton Station. *(H C Casserley)*

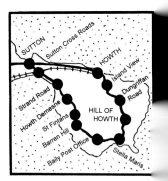

No 4 at Hill of Howth on August 1958. *(J C W Halliday*

Railcar F enters the branch platform at Howth Junction in August 1957. On dissolution of the GN, it became UTA No 104 and survived until closure of the Warrenpoint branch in 1965. The Howth branch has been electrified since 23 July 1984 and the main line as far as Malahide since 9 October 2000. *(Peter Griffin)*

n experience best enjoyed in favourable weather. A tram pulls away from Howth where the terminus was at the Dublin end f the station. A steam loco is running round its train. The signal box closed on electrification in 1984. *(J C W Halliday)*

The Belfast Central Railway

"A" class 0 – 6 – 0 No 150 shunting Donegall Quay in December 1956. This loco was retained for use on the Quay, as it could cope with the curves and restricted clearances.   *(Peter Sunderland)*

From 1848, Belfast had three main line termini. None of them were convenient for the City Centre or the Docks, nor was there any rail connection between them. It was to remedy this situation that the Belfast Central Railway was authorised in 1864. The scheme promised a Central Station, linked to all three Railways.

The Main Line Companies showed little interest. The project was financed from London. Apart from a short lived passenger service, its achievement was a goods link between the Great Northern and County Down Railways and the Docks. It built neither a Central Station nor a direct connection to the Northern Counties Railway. Belfast retained its three termini until 1976. Only in 1994 were the three Railways connected for passenger trains.

The Belfast Central began at Central Junction, half a mile outside Great Victoria Street. The double track main line ran for two miles to Queens Bridge. A mile before the terminus, at East Bridge Junction, a branch diverged sharply to the right, crossing the Lagan by an eight span single track viaduct to join the Belfast & County Down Railway at Ballymacarret Junction.

The line was virtually complete by the end of 1870 but did not open until 1875 whilst arguments took place over the layout of junctions with neighbouring Companies and about running powers. A passenger service was provided of up to 18 trains a day between a platform at Central Junction and Queens Bridge. There were intermediate stations at Botanic and Ormeau but no interchange with any other trains.

In 1879, a short single track extension was opened from Queens Bridge to join the Belfast Harbour Commissioners lines on Donegall Quay. By this route, the Belfast Central could reach the cross channel steamer berths and the Belfast & Northern Counties Railway but only via the dockside lines with their sharp curvature. The descent from Queens Bridge involved a 150 yard tunnel with restrictive clearances.

The Belfast Central was taken over by the Great Northern in August 1885. Almost at once, the passenger service was withdrawn at the end of November. Four locomotives passed into G N stock. One of them, a Beyer Peacock 2 – 4 – 0T of 1880 survived as G N No 195 until 1950.

Four 0 – 6 –4 Ts of class RT, Nos 22,23,166 and 167 were built between 1908 and 1911. They had the profile to work through to Donegall Quay.

Traffic over the Central included regular livestock trains through to the ships on Donegall Quay. Transfer freights worked from Great Victoria Street bringing wagons, which had sometimes worked in attached to passenger trains, with goods for

Donegall Quay.

Coal for both railway use and general sale was discharged from colliers at the South Quays. The Great Northern had a headshunt at Ballymacarret Junction from which it could gain access to the Harbour Commissioners lines on this side of the Lagan. Oil was also brought in by sea to the South Quays.

The Belfast Central line was worked by permissive block except when passenger or oil tank trains ran when special arrangements had to be made to enforce absolute block. Passenger trains were mostly excursions between Great Northern stations and Bangor. Early in 1963, one track was re-laid between Central and East Bridge Junctions. It then became the practice, when passenger trains ran, to institute single line working on the good track.

Three of the class RT locos passed to the UTA on break up of the Great Northern in 1958. 22 had just been scrapped. 23 retained its number. 166 and 167 became 24 and 25. In the late 1950s, Queens Bridge Tunnel was deepened to give clearance for a greater variety of locomotives, vans and, potentially most important, containers. The two ex Sligo, Leitrim & Northern Counties locos, Nos 26 "Lough Melvin" and 27 "Lough Erne" became regular Belfast Central engines, along with Northern Counties 0 – 6 – 0 No 13.

It was all too late. The East Bridge Junction to Donegall Quay section was closed at the end of May 1963 to facilitate work on the new Queen Elizabeth road bridge. Shortly afterwards, Maysfields goods yard was closed and the route from Central Junction to Ballymacarret reduced to one single line section.

The UTA ceased to carry general freight in February 1965. Passenger excursions ran over the Belfast Central until July. 0 – 6 – 0 No 49 (ex GN No149) had eight coaches on the last excursion from Lisburn to Bangor and back on Wednesday 28 July. Two days later, a four coach DMU worked an excursion from Bangor to Dublin and back. Soon after that, a bridge was removed to permit road widening.

The subsequent revival of the Belfast Central will feature in *Part Two* as part of the story of Northern Ireland Railways.

4 – 4 – 0 No 66 "Meath" running light past Central Junction in August 1959. Great Victoria Street is behind the camera, Adelaide straight ahead and the Belfast Central line behind the signal box. The third track, on the extreme right, was available for light engine movements from Adelaide to Great Victoria Street. *(Peter Sunderland)*

The bridge carrying the Belfast Central Railway over the Lagan was known colloquially as the "Shaky Bridge". This 1960 view is from the County Down side, looking towards the site of the present day Central Station. The bridge was replaced by a sturdier structure before the line reopened in 1976.
(Peter Sunderland)

Great Northern Motive Power

198 "Lough Swilly" entering Enniskillen with a stopping train from Dundalk in 1955.

(Peter Sunderland)

At the amalgamation in 1876, the Great Northern inherited 118 steam locomotives. These were maintained at five workshops under four locomotive superintendents. A decision was made to concentrate on a new works at Dundalk, for which land was purchased in 1880. Dundalk was already home to the Irish North works at Barrack Street and to the smaller Dublin & Belfast Junction shops alongside the main line. It was also the mid point between Dublin and Belfast. By 1885, the locomotive department was under unified management at Dundalk.

Over the period 1876 to 1948, the Great Northern acquired 277 locomotives. 47 of these were built at Dundalk. The remainder came from outside suppliers of whom Beyer Peacock was by far the most popular with 152 examples.

In 1930 there were still seven locomotives in service, which predated the 1876 amalgamation. The last of these were not withdrawn until 1948 by which time they were up to 77 years old. Only 25 new locomotives were built after 1930, plus eight almost complete rebuilds. The age of the fleet was rising.

Five class V passenger engines came from Beyer Peacock in 1932. These were numbered 83 to 87 and named after birds of prey; "Eagle", "Falcon", "Merlin", "Peregrine" and "Kestrel". They were

4 – 4 – 0 three cylinder compounds, designed to exploit the rebuilding of the Boyne Viaduct, which had raised the maximum axle load from 17 to 21 tons. Five 0 – 6 – 0 goods engines, Nos 78 to 82, class UG, were built at Dundalk in 1937. The GNR Works then turned its attention to eight 4 – 4 – 0s of class S, which had originally come from Beyer Peacock in 1913 – 15. They were rebuilt with heavier frames. Nos 170 to 174 and 190 to 192 were given names of mountains; "Errigal", Slieve Gullion", "Slieve Donard", "Galtee More", Carrantuohill", "Lugnaquilla", Croagh Patrick" and Slievenamon"

The Second World War delayed further new building until 1948, when the GNR received its final 15 steam locos from Beyer Peacock. 145 to 149 were five more class UG 0 – 6 – 0s. 201 to 205 were class U 4 – 4 – 0s, built to a design of 1915 and named after counties; "Meath", "Louth", "Armagh", "Antrim" and "Down". Their 33 year old class mates, 196 to 200 were then given names of Loughs; "Gill" "Neagh", " Swilly", "Derg"and "Melvin". Last came the class VS 4 – 4 – 0s, Nos 206 to 210, named after rivers; "Liffey", "Boyne", "Lagan", "Foyle" and "Erne". These were similar to class V but were simples not compounds. They were distinctive in appearance thanks to smoke deflectors.

At the dissolution of the Great Northern, on 30 September 1958, 166 steam locomotives were

Starting with the J class in 1885 (see "Glencar" on page 75) 4 – 4 – 0 became the norm for G N passenger locos. U class 202 "Louth", built 1948, stands at Bundoran Junction with a train from Bundoran to Enniskillen on 17 September 1955.

(John Oxley)

For short distance work such as Dublin to Howth, they had some 4 – 4 – 2 tank engines, the extra trailing wheel being needed to support the bunker. Class T2 No 62 (Beyer Peacock, 1929) arrives at Amiens Street on 10 September 1955. *(John Oxley)*

For goods engines, the GN stuck to the 0 – 6 – 0 arrangement so that the full weight was on the driving wheels. 179 (Beyer Peacock, 1913) is heading north from Dublin Amiens Street with a Guinness train on 10 September 1955. *(John Oxley)*

divided between CIE and the UTA. As far as possible, each individual class was divided equally. CIE kept the GN numbers with an "N" prefix. The UTA applied new numbers.

All former GN locos were withdrawn by 1965, giving way either to diesels, to further line closures, to the abandonment of goods traffic or, within Northern Ireland, to ex NCC steam locomotives. Even before 1958, the UTA had loaned a few of its newer engines to supplement the ageing GN fleet.

On 29 October 1966, the UTA marked the end of steam on the ex GN with an excursion from Belfast to Dublin and back. This was worked by ex NCC 2 – 6 – 4T No 54 hauling eight GN coaches. The following Saturday, 5 November, sister engine No 51 worked an eight coach football special from Belfast to Portadown and back. Adelaide shed closed that weekend.

Preservation
Four GN steam locomotives have been preserved:

JT	2 – 4 – 2T	93	
S	4 – 4 – 0	171	"Slieve Gullion"
V	4 – 4 – 0	85	"Merlin"
Q	4 – 4 – 0	131	

The JT is at the Ulster Folk & Transport Museum, Cultra. The three 4 – 4 – 0s are in the care of the Railway Preservation Society of Ireland at Whitehead, from where they can operate excursions over both Northern Ireland Railways and Iarnrod Eireann.

Great Northern Diesels
From 1926, the County Donegal Joint Committee had experimented with petrol driven railcars. By 1931, it was operating two diesels, the earliest such vehicles in the British Isles. The narrow gauge County Donegal enjoyed technical and engineering support from the Great Northern, one of its joint owners. The parent Company observed progress with a view to introducing the technology on its own system.

The Great Northern put two diesel railcars into service in 1932. "A" had mechanical transmission, "B" electrical. Both seated 32 passengers, were carried on bogies and could be driven from either end.

Two years later came railcar "C" (later C1). This had the power unit mounted on one bogie with the passenger body articulated on. It could be driven from one end only. The following year, "C2" and "C3" emerged as a pair coupled back to back. They had no multiple unit control. The leading car ran under power. The trailing one was pulled along in neutral.

A major step forward came with "D" and "E" in 1936, followed by "F" and "G" in 1938. Each of these had an outer coach articulated onto a central power unit. "F" and "G" could each seat 164 passengers and could attain a speed of 48mph. Streamlined in appearance, they were revolutionary for their time.

In search of something cheap for lightly used branch lines, six "railbuses" were built in 1934/5. Essentially, these were road buses converted to run on rails. They retained their pneumatic tyres around

Railcar G, built 1938, at Howth in Summer 1958. On break up of the GNR, it went to CIE but was sold to the UTA in 1961, becoming No 105. It was withdrawn on closure of the Warrenpoint branch in January 1965. *(Peter Sunderland)*

4 – 4 – 0T No 96 in use as a pile driver at Ballyhaise in 1920, shortly before scrapping. This was one of the first three class BT locos, supplied by Beyer Peacock in 1885 for use on locals out of Belfast. They were named "Lisburn", "Balmoral" and "Dunmurry". Ten more were built at Dundalk over the next few years but all were withdrawn around 1920.
(LCGB – Ken Nunn Collection)

No 74 coaling at Londonderry in July 1957. 17 class PP 4 - 4 - 0s were built between 1896 and 1911. No 74 was supplied by Beyer Peacock in 1896. It carried the name "Rostrevor" until about 1920. The rail mounted coaling plant appears also to be of some vintage.
(Peter Sunderland)

One of the BUT railcars being assembled at Dundalk Works in 1957. *(Peter Sunderland)*

125, 20, 118, 77, 157 and 126 at Belfast Shed on 17 September 1909. The shed was located on the north west side of the line, just outside Great Victoria Street. It closed in March 1911, in favour of a new facility at Adelaide, allowing the goods depot and carriage sidings to expand.

(LCGB – Ken Nunn collection)

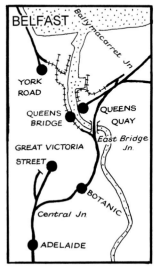

Looking across the shed yard at Adelaide in May 1956 with 208 "Lagan", 209 "Foyle" and 83 "Eagle ", all resplendent in Great Northern blue livery.

(Peter Sunderland)

0 – 6 – 0T No 31 at Dundalk Works in 1957. The only example of a crane loco on the GN was supplied by Hawthorn Leslie of Newcastle in 1927

(Peter Sunderland)

which a steel rim was fitted – the Howden – Meredith patent that had previously been tested on road lorries adapted as permanent way vehicles. Two of the railbuses worked on the Dundalk, Newry & Greenore and one was sold to the Sligo, Leitrim & Northern Counties Railway.

The Second World War delayed further development until 1948 when an order for 20 vehicles was placed with A E C, Southall who had built railcars for the (English) Great Western Railway. Delivered in 1950/51 and numbered 600 – 619, they were formed into 10 three car multiple units by the insertion of an adapted ordinary coach between each pair of power cars. Two centre cars were possible on lines with modest gradients.

At this stage the Great Northern was ahead of any railway in the British Isles in the operation of diesel passenger trains. In March 1954, the GN Board sought Government approval for 24 more power cars. Authority was given by the end of the year and an order placed with British United

Traction, which had taken over AEC. Nos 701 to 716 were second class power cars with driving cabs and gangway connections at both ends. 901 to 908 were composites with a driving cab at one end and gangway at the other. They were intended to work in a variety of formations, strengthened by ordinary coaches adapted for multiple unit working.

The first three BUT cars did not enter service until June 1957. By the time delivery was complete, the Great Northern had ceased to exist and the last of these BUT cars were allocated to CIE or the UTA on arrival. British United Traction had taken over AEC.

In 1954, the Board also invited tenders for diesel locomotives in three power ranges: 350 to 400hp, 800hp and 1,000hp. Machinenbau A G of Kiel, Germany were sufficiently keen to send an 800hp loco at their own expense for trials on the GN and UTA. The GN bought it and gave it the number 800 but the Government of Northern Ireland blocked any further purchases.

The unique MAK diesel shunting passenger stock at Belfast Great Victoria Street during 1955. It became GNR No 800, then CIE No 801 but was withdrawn from traffic in 1967. It was reinstated very briefly in 1974 as station pilot at Drogheda. After that it was sold to a Galway scrap merchant for use as a stationary generator. Meanwhile, similar locomotives did find widespread favour in Germany, where derivations of the MAK design are still in use.
(Peter Sunderland)

A driving trailer, converted from an ordinary coach, leading a diesel railcar out of Antrim with a local for Belfast via Lisburn in August 1959.
(Peter Sunderland)

General Motors

One of the invitations to tender was sent to General Motors. Protocol determined that the response should be sent via a garage in Dublin, which held the General Motors franchise for Ireland.

The tender document, now in the hands of Peter Sunderland, proposed that standard GM products, their G6 and G8 models, could be modified to meet the requirements of the GN. Evidence was offered as to the reliability of the locos and their low repair costs. If a deal were signed, GN supervisory staff would be invited to La Grange, Illinois for free tuition on operation and maintenance of the locos but the GN would have to pay for their travel and accommodation. After delivery, General Motors would pay first class fares for one or more of their representatives to superintend putting the locos into service. They would stay to be consulted for an agreed period.

The price was around £48,000 per loco "free on rail" at London, Ontario, Canada. Shipping costs, estimated by General Motors, were just over £3,000 per loco. This equates to more than £1m per unit at 2005 prices. More relevant, it compares with £29,500, which the G N actually paid for the solitary diesel acquired from MAK. We don't know what MAK had tendered for new locos. £29,500 may have been a knock down price for an ex demo model, which would otherwise have had to go back to Germany at MAK's expense.

Politics ensured that nothing came of the GN's quest for further diesel locos. Its southern neighbour, CIE did gain Government approval to invest in both railcars and locomotives. For the latter, it put its faith, initially, in Metropolitan Vickers. But, from the early 1960s, it turned to General Motors and today, every locomotive on both Iarnrod Eireann and Northern Ireland Railways is from General Motors. So are all the post privatisation locos on English Welsh & Scottish Railway.

The Great Northern had taken the lead with diesel railcars. In a different political climate, it might also have led the way with locomotives. It could have been the first to see the merit of using well tried American technology, when British diesels were still in the experimental stage.

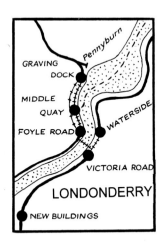

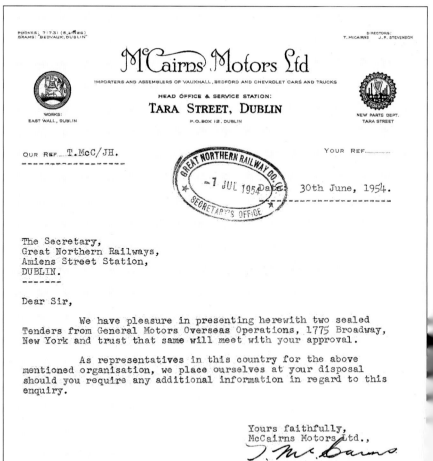

Whitsuntide 1963

The Special Traffic Notice for 1 to 7 June shows that Sunday school excursions were still an important consideration.

On Whit Saturday 1 June, the 8.15am Belfast to Dublin Express had a special stop at Skerries to set down Mrs Galbraith's party of 10 people. There were special trains to Dublin at 8.10am from Lisburn for a party of 120 and 8.55 from Belfast, for 250 people, both returning in the evening.

Great Victoria Street Baptists and Windsor Presbyterians combined to offer 710 passengers for a special to Portrush. Leaving Belfast at 9am and calling at Adelaide, the train went via the Knockmore to Antrim branch, which normally saw only one goods train per day and which was subject to an overall 25 mph speed restriction. Through journey time to Portrush was 2 hours. They returned in the evening.

No fewer than 3,650 passengers were booked on a total of six excursions, two from Dungannon, two from Portadown and two from Lisburn which made their way to Bangor via the normally freight only Belfast Central line. Pilotman working was in force on the Belfast Central, which was then only worked on one track. There were a lot of light engine and empty carriage movements during the day, prior to the evening departures from Bangor at 6.20, 6.45, 7.05, 7.35, 8.05 and 8.45.

On Sunday 2 June, Mrs Galbraith's party came home, courtesy of a special stop at Skerries by the 6.20pm ex Dublin.

The Down & Connor Pilgrimage to Claremorris, County Mayo, for the holy shrine at Knock, required ten coaches "including Kitchen Car and Ambulance Coach". Goodness knows what they were intent on getting up to but, lest it involved contraband, the train was to be Customs examined prior to its 6am departure from Belfast, on arrival at Dublin and again on arrival back in Belfast at 12.35am. It was ten years since the Irish authorities had given up worrying what people took out of their territory but there was still no corresponding relaxation on the British side.

Half day excursions to Warrenpoint were offered by a special DMU at 2pm from Portadown, also by extending the regular 2pm Belfast – Newry DMU.

Normally there were no Sunday trains to Warrenpoint, where arrivals were at 2.50 and 3.28. These were advertised excursions, for the general public, not sold out to any particular group. There was a return train from Warrenpoint at 5.45pm, only to Newry with no onward connection. The main return train was at 8.15 and here there were two contingencies, depending on the volume of traffic. If the load was modest, then on arrival at Newry at 8.30, the DMU would wait ten minutes and become the normal 8.40 to Belfast. If the train was full from Warrenpoint, it would work through to Belfast independently, in advance of the 8.40. In this event, passengers from Warrenpoint for Poyntzpass and Tanderagee must be told to change at Newry because the special would not be stopping. The decision would have to be made as soon as they had gauged the load on the outward journey because it determined which of the two DMUs did which parts of the roster.

On Thursday 6 June, the Christian Brothers were on the move for their Educational Tour of Dublin. Numbers are not quoted but the Special at 7.25am from Londonderry had to be double headed as far as Portadown. As the double heading had to be arranged by Londonderry Shed, the empty stock movement, at 12.55am from Belfast was presumably single headed. Return was at 7.25pm from Dublin with arrival at Strabane 11.14 and Londonderry Foyle Road 11.45pm. Some modifications were required to goods train workings to accommodate this Special. No such problems surrounded the running, the same day of a Dublin to Belfast advertised excursion. This simply followed a conditional path in the working timetable, reserved for such an occasion, at 8.45am from Dublin and 5.50pm from Belfast.

Arrangements like this continued through this and the following summer. Two years later, there was no "Derry Road", no Warrenpoint Branch and no Belfast Central, so no way of getting to Bangor. Even on surviving routes, the days of the Sunday school excursion were coming to an end, victim of changing social patterns and a more rigorous costing of special trains

The Sligo, Leitrim & Northern Counties Railway

Railcar 2A with luggage trailer in the Sligo bay at Enniskillen in 1957. Visible are the Howden – Meredith wheels (steel flanges around the pneumatic tyres), crude brake blocks and starting handle.

(Peter Sunderland)

For 75 years, Sligo enjoyed a rail outlet towards Belfast and Londonderry courtesy of this 5ft 3in gauge line, which extended 42 miles from Carrignagat Junction, near Sligo to Enniskillen where it connected with the Great Northern Railway. Always impoverished, the SL&NC spent time in receivership yet managed to retain its independence throughout its existence.

The Railway was authorised by an Act of August 1875. Anticipating difficulty in raising capital, the promoters had sought to include provision for Baronial guarantees. These would allow dividends to be made good out of local rates if the fortunes of the Railway itself could not support them. This part of the measure failed to get through Parliament so, instead, guarantees were given by some of the promoters/directors themselves.

There was difficulty finding a contractor willing to build the line at a price the Company could afford. In March 1877, the Chairman, Arthur Loftus Tottenham of Glenfarne Hall and the Engineer, Mr Barry, resigned their positions and set up in partnership as building contractors. Work was then able to start from the Enniskillen end.

The line opened as far as Belcoo on 12 February 1879 for goods traffic only. Passengers had to wait until 18 March after an inspection by Major General Hutchinson of the Board of Trade.

Progress beyond Belcoo was possible only with the aid of a loan from the Irish Board of Works, interest on which had to be guaranteed by some of the directors and shareholders. Opening to Glenfarne was achieved on 1 January and to Manorhamilton on 1 December 1880. The Company was observing the convention that new facilities should open on the first day of a month so that *Bradshaw* might be kept as accurate as possible.

The next section to Collooney opened on 1 September 1881. Here there was already a station close by on the Midland Great Western Railway but the SL&NC still had to build another 1¼ miles of line to Carrignagat Junction and then it had to pay the cost of doubling the MGWR for nearly a mile to Ballysodare Station. The SL&NC tried to negotiate its way out of this latter requirement but the MGWR would not move.

The SL&NC struggled on to complete these works and the final stretch opened on 7 October 1882 to Carrignagat Junction. From here the SL&NC exercised running powers over the MGWR both to Sligo Station and along the Quay branch. A service of three trains was offered in each direction

4 – 4 – 0 "Glencar" was built in 1887 as Great Northern J class No119 "Thistle". It was bought by the SL&NC in 1921 but only ran until 1928, when it was cannibalised to keep sister engine "Blacklion" (GN No 118 "Rose") going another three years. The name "Glencar" was transferred to GN 0 – 6 – 0 No 31, built 1890, which ran on the SL&NC from 1928 until 1949.
(Peter Sunderland collection)

0 – 6 – 4T "Lissadell" at Manorhamilton in 1957. Built in 1899 by Beyer Peacock, the engine was sold to a scrap merchant in 1954 but it remained at Manorhamilton until closure of the Railway. Belfast Transport Museum made an unsuccessful bid for sister engine "Hazlewood", which had remained in service to the end but it went to the same scrap man.
(Peter Sunderland)

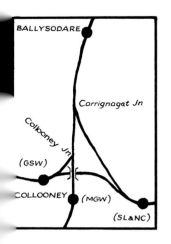

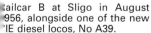

Railcar B at Sligo in August 1956, alongside one of the new IE diesel locos, No A39.
(Peter Sunderland)

75

Sundays excepted. This was the basic passenger timetable for most of the following 75 years. Some of the passenger trains ran as "mixed" conveying goods traffic as well.

Mr Barry had pulled out of the contracting partnership in 1879 leaving Mr Tottenham to carry the financial risk alone. He had to accept part payment in shares. Two locomotives were ordered from the Avonside Engine Company but supplied via the Bristol Wagon Company who built the first carriages and wagons. The SL&NC could not afford to pay for them. During 1882, the debt was taken over by Charles Morrison, the largest shareholder in the SL&NC, who then leased the equipment to the Railway. In 1886 the company tried to borrow further from the Irish Board of Works in order to pay off Mr Morrison but were unsuccessful.

The Company fell into arrears with interest on its existing loan, causing the Board of Works to apply to the Court both for the appointment of a receiver and for an enquiry into the Railway's working and management. As was quite common at the time, the Secretary and Manager, R E Davis, simply became Receiver and Manager. E J Cotton of the Belfast & Northern Counties, who was also Chairman of the Irish Railway Managers Conference, conducted the Enquiry. His Report identified some inefficiencies but the main problem was the low level of traffic potential.

The Board of Works, as principal creditor, were thought to favour sale of the line to the Great Northern and Midland Great Western, either jointly or dividing it at Manorhamilton. There was considerable opposition to this from traders who feared that the two large Companies would fail to develop the SL&NC as a through route.

The SL&NC placed a lot of faith in the forthcoming opening of the Claremorris to Collooney line on 1 October 1895. This was worked by the Waterford & Limerick Railway, which soon changed its name to Waterford, Limerick & Western. It provided the village of Collooney with its third station before joining the MGWR at Collooney Junction whence it had running powers into Sligo. It also threw off a half mile spur which passed underneath the MGWR to join the SL&NC at the entrance to its own Collooney Station. The junction here was paid for by the WL&W whose successors continued to pay a small rent to the SL&NC right up to 1956 when the direct spur was abandoned.

Only special goods trains travelled direct between the WL&W and the SL&NC. Passenger and ordinary goods traffic off both lines worked into Sligo. If the Claremorris link was the salvation of the SL&NC, it was not for the volume of business it produced. Unfortunately, the West and South West were not fertile sources of traffic. What the Claremorris line did achieve was point the way to a

"Lough Erne" shunting the SL&NC yard at Enniskillen in 1957. The Company Offices are on the platform. The GN South Cabin occupies an elevated position.   *(Peter Sunderland)*

"Lough Erne" shunting cattle wagons and passenger stock, in preparation for the 7.20pm "mixed" departure from Enniskillen. Railcar B can just be seen in the bay platform. *(Peter Sunderland)*

refinancing of the SL&NC.

The Claremorris to Collooney line was built with the help of Baronial guarantees and a grant from the Irish Board of Works. These facilities had become much easier following the Tramways & Public Companies (Ireland) Act, 1883 and the Light Railways (Ireland) Act 1889. The SL&NC's grievance was that it could have benefited from these measures had it not been built too soon. Had its promoters foreseen the changes in legislation, they could have delayed their investment. They felt that they were now being penalised for their courage in going ahead when they did.

In 1895, the SL&NC put its case to the Government but had to wait two years for an answer. Possibly the Government hoped that traffic off the Claremorris line would ease the problems of the SL&NC without the need for any concession. This cure did not materialise to anything like the extent hoped. In 1897 the Board of Works accepted a lower rate of loan interest, with retrospective effect, on condition that arrears were cleared in priority to the claims of Mr Morrison for overdue rolling stock rental. The Receiving Order was then lifted, Mr Davies returned to his former position and the Company gained ownership of its rolling stock.

The Railway then settled down to a period of stability. Employing upward of 130 people, the line served a declining population in an area with little industry. In 1871, the combined population of Counties Sligo and Leitrim stood at 211,000. This figure was only 60% that of 1841, before the Famine. By 1956, the last full year of the Railway, it was down to 94,000.

In 1921, the Sligo, Leitrim & Northern Counties gained international status. Two years later, Customs posts were established on Glenfarne and Belcoo Stations. Eventually, agreements were made for Irish Free State goods to travel through Northern Ireland in sealed bonded containers. Likewise, traffic from Great Britain to Northern Ireland could pass through the port of Sligo.

The Border had a detrimental effect on the Railway as Sligo began to look exclusively towards Dublin as its commercial capital when, previously, this position might have been shared with Belfast. The situation deteriorated during the 1930s when a "tariff war" broke out between the Irish Free State and the United Kingdom. From 1935, the Government of Northern Ireland paid a subsidy to the SL&NC whose operating loss it attributed to the actions of the Dublin administration.

The subsidy agreement was badly worded so that the Government undertook to match the Company's profit or loss. It was obliged either to make good the Company's loss or to double up its profit - if the Railway ever made one. This may have seemed impossible but it actually happened during the Second World War.

Diesels

The SL&NC was desperate to find a cheaper way of operating the service. In 1932, it borrowed one of the Great Northern Railway's new diesel railcars for a trial run from Enniskillen to Sligo and back. Two years later, the Company ordered an even lighter vehicle of its own. Entering service in June 1935, railcar "A" was a former Great Northern petrol bus, fitted with "Howden – Meredith" steel tyres and partnered by a two ton luggage trailer. The petrol engine was replaced by a Gardner diesel in 1938 but the vehicle was destroyed in an accident at Glenfarne on 7 March 1939.

Meanwhile, railcar "2A" had arrived in April 1938. This was another former Great Northern bus, re-equipped with Gardner engine and "Howden – Meredith" wheels. From June 1938, the normal passenger timetable was rostered for the two railcars but the loss of "A" caused a return to steam on some workings. A replacement "A" appeared after a few months in the form of Great Northern railcar "D1". It was given the nearly new Gardner engine from the previous railcar "A". By 1950, the body was beyond repair so it was sent to Dundalk for a new one, also recovered from a Great Northern bus.

Ordered in 1944 and delivered in 1947, railcar "B" was a Gardner – Walker product on the same principal as Great Northern "C1", "C2" and "C3" with the passenger section articulated onto the power unit. It could be driven from either end.

The Second World War and After

The War gave the Company a breath of life. When the Great Southern Railway was reduced to running on only two days a week, Sligo to Dublin traffic was able to go via Enniskillen because cross border railways had better supplies of fuel. The SL&NC earned modest profits to which the Northern Ireland subsidy was added. It used this new found wealth to purchase railcar "B".

In 1947, the Railway placed an order with Beyer Peacock for two new 0-6-4 tank engines. These turned out to be the last new steam locomotives delivered to an Irish railway, a distinction which would not have been earned had they been shipped on completion in 1949. Unfortunately, the Company could not afford to pay for them. There was a two year delay whilst a hire purchase deal was agreed involving a guarantee by the Government of Northern Ireland.

From 1952, the Railway received a subsidy from the Republic of Ireland but Northern Ireland threatened to withdraw their's from the end of 1955. The Company gave notice to staff but the threat of closure on 31 December 1955 was lifted by an increased subsidy from the Republic and a levy on cattle market sales.

The final obstacle, which the SL&NC could not overcome, was the closure of the Great Northern Railway through Enniskillen on 30 September 1957. The Republic announced that it could not continue to subsidise the Sligo line beyond that date because

0 – 6 – 4T "Sir Henry" at Manorhamilton with the 3.45pm "mixed" from Sligo to Enniskillen on 19 May 1924.
(LCGB - Ken Nunn collection)

the absence of an onward link would render it futile. As late as 6 September, the Northern Government told the Ulster Farmers' Union that it would not retain even a goods link from Omagh to Enniskillen because the Republic was going to force closure of the SL&NC. On 9 September, the Republic agreed to continue support provided there was at least a goods outlet from Enniskillen to Omagh. On 18 September, Northern Ireland refused any concession on the Great Northern closure. Two days later, the SL&NC announced that it would cease all activity, including its road operation in the Republic. It gave staff two weeks notice, the first four days of October being used to work empty "foreign" wagons off the line.

Railcar "B" was sold to CIE. The two newest steam locomotives, "Lough Erne" and "Lough Melvin", still the property of Beyer Peacock, were bought by the Ulster Transport Authority.

The Last SL&NC Timetable

Covering the period from 25 June 1957 "until further notice", the booklet runs to no fewer than 32 pages. Many of these are given over to adverts, both internal and commercial. The middle two pages show the basic service of two trains each way, plus a third from 24 June to 7 September, plus the unbalanced steam train at 7.20pm from Enniskillen. Five intermediate stations are shown with no trains stopping, but a footnote indicates that railcars will stop at all halts on request. The one steam train is shown by footnote to stop only at certain halts by request.

Two pages are devoted to connections from SL&NC stations to destinations on the Great Northern Railway. Travel is not fast. By the 6.20am from Sligo, passengers can be in Londonderry Foyle Road at 11.35am. 100 miles in 5¼ hours, of which 50 minutes is spent changing at Enniskillen and 19 at Omagh. Both have refreshment rooms. Belfast Great Victoria Street (134 miles) is reached at 12.55, with the same two changes and a restaurant car on the last leg of the journey.

Possibly more optimistic is the thought of passengers booking in any number from SL&NC stations to London, Birmingham, Liverpool or Glasgow. None the less, seven pages are devoted to cross channel connections, one for each of the principal sea routes; Dun Laoghaire – Holyhead, Dublin – Liverpool, Belfast – Liverpool, Belfast – Heysham, Belfast - Glasgow, Larne – Stranraer and Londonderry – Glasgow. Many of the connections are appalling enough, using the 9.30am from Sligo but this train only runs 24 June to 7 September. Outside this period, in all seven cases, passengers have to leave 3 hours, 10 minutes earlier, just to spend more time en route.

The above connections are all via Enniskillen. Those available with CIE at Ballysodare are mentioned only as a footnote to the main table.

A brave effort. It wasn't the Railway's fault that it served an area of such sparse population or that its neighbours would never take it over

he two 1949 built 0 – 6 – 4T locos pass at Glenfarne in September 1957.   *(J C W Halliday)*

The largest engineering structure on the SL&NC was Weir's Bridge, which crossed the Erne a mile from Enniskillen. The Railway could never afford to maintain the bridge properly and, after 1906, it was subject to a speed limit of 5 mph with a ban on double heading. The view is towards Enniskillen.

(J C W Halliday)

Railcar B, undergoing customs examination at Glenfarne in August 1956. The view is towards Sligo. As CIE No 2509, the railcar was withdrawn from service in 1971 but not scrapped. In November 2005, it arrived in very poor condition, at the Downpatrick & County Down Railway, for eventual restoration.

(Peter Sunderland)

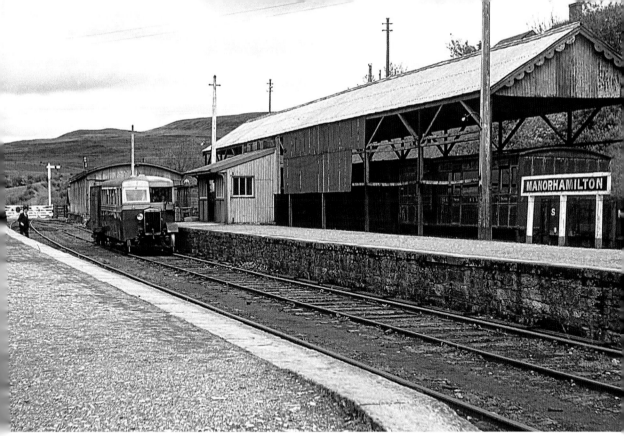

SL&NC Railcar A at Manorhamilton on an Enniskillen to Sligo working in September 1957. *(John Carter)*

Great Northern Railcar C2 on the turntable at Cavan, having worked the daily parcels from Dundalk on 2 August 1958. *(John Carter)*

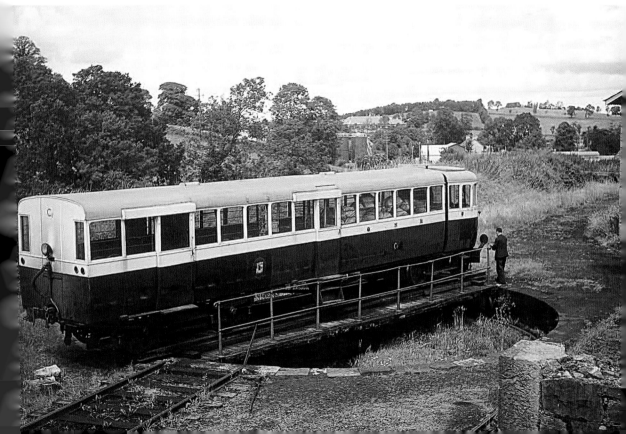

The Londonderry & Lough Swilly Railway

4 – 8 – 0 No 12 near Pennyburn with the 8.20am from Burtonport to Londonderry on 8 May 1920.
(LCGB – Ken Nunn collection)

The northern part of County Donegal offered little inducement to railway promoters. The Lough Swilly Company struggled to achieve a 30½ mile network, reducing its gauge in the process. Then, at the beginning of the twentieth century, the Government stepped in to finance extensions which took the route mileage up to 98½.

Operating through some of the wildest and most scenic parts of the British Isles, the Lough Swilly ran the largest engines on the narrow gauge. But its supremacy was short lived. By 1931, it was decided that the Railway should close as soon as the road network became adequate to carry the traffic. The process was slowed down by the fuel shortages of the Second World War, when the Railway experienced a partial revival. It finally succumbed in 1953, since when the Company has continued as a road transport undertaking.

Origins
Plans for a standard (5ft 3in) gauge Londonderry & Lough Swilly Railway were deposited in Parliament in 1852. The line was to be 8¾ miles in length and was to utilise the two mile Trady Embankment, which had been completed in 1850 to keep the tide off 2,000 acres of reclaimed land.

Beyond the Embankment, which stretches from Tooban to Farland, the railway was to curve out onto a pier at Farland Point, whence it was hoped that steamships would provide onward connections.

The Bill passed in June 1853, but no progress was made and a second Act had to be obtained in 1859. This time a contractor was engaged and the line was completed in time for opening on the last day of 1863. Two trains ran in each direction each weekday at 7.30am and 4pm from Londonderry. A steamer called "Swilly" plied in connection to points on the far side of the Lough.

Buncrana replaces Farland
Powers were obtained in 1861 for a branch from Tooban to Buncrana. Work on this proceeded quickly and the line opened in September 1864.

Soon it was decided that the line to Farland Point had been a mistake and, during 1865, the train service to there was suspended for want of steamer connections. It resumed briefly when a steamer was arranged, only to stop permanently in July 1866. The pier at Farland Point was dismantled and rebuilt at Fahan, an intermediate station on the Buncrana line.

The distance from Londonderry to Buncrana was 12 miles. Gradients were fairly easy but the route was scenic, particularly on the coastal stretch between Inch Road and Buncrana.

Illegal Operations
The terminus in Londonderry was at the Graving Dock, 1½ miles north of the Great Northern station

This page top Kilmacrenan Station and goods yard, looking towards Burtonport in 1937.
(H C Casserley)

Middle Owencarrow Viaduct, looking towards Burtonport in 1937. *(H C Casserley)*

Bottom Middle Quay, Londonderry with a Fleetwood Steamer alongside. The Harbour Commissioners' mixed gauge track extended from the Swilly Station at Graving Dock to the GN at Foyle Road and across the Carlisle Bridge. The Lancashire & Yorkshire/ London & North Western Joint Railway steamer ran twice a week between 1904 and 1912.

Page 85 upper The sole facility at Cornabrone Halt, part way between Ballyduff and Kiltubrid, was a nameboard in the hedge on the far side of the road. 3T pauses with the daily "mixed" from Ballinamore to Arigna. *(J C W Halliday)*

Lower Kiltubrid boasted a house and a waiting shelter on the curved platform, also a goods siding. 3T calls there on the return journey from Arigna. *(J C W Halliday)*

Next page upper 4 – 4 – 0 No 207 "Boyne" passing Adelaide in May 1964 with a relief for the Belfast to Dublin "Enterprise Express" *(Peter Sunderland)*

Lower 2 – 6 – 0T No 4T at Ballinamore in July 1957, waiting to depart for Dromod. The passenger coach was rebuilt at Inchicore in 1951 out of bus body parts. The roof leaked, there was no heating and lighting was rationed because the batteries had to be sent to Dublin for recharging. *(Peter Sunderland)*

at Foyle Road. Connecting the two were the dockside lines of the Londonderry Port & Harbour Commissioners.

The 1859 Londonderry & Lough Swilly Railway Act forbad the use of steam engines over Strand Road level crossing which lay about 200 yards north of Graving Dock Station. From time to time, the Board of Trade protested at the flouting of this law but the Railway Company took no notice. It was not practical to bring full trains into the terminus by horse. An Act of 1918 finally repealed the ban on steam working into Graving Dock Station, thus recognising what had been going on for 55 years.

Graving Dock was hardly convenient as a City terminus so, from 1 January 1869, Lough Swilly trains by-passed their own terminal station and ran over the Harbour Commissioners' line for a mile to Middle Quay, where a passenger station was built, just ½ mile short of the Great Northern Railway.

This arrangement was also illegal as the dockside line was not passed to carry passenger traffic. It was a fall out over money, not the law, which stopped the practice at the end of 1884. Having failed to extract an increased rent from the Lough Swilly, the Harbour Commissioners removed a few rails near the boundary of their property, forcing trains to terminate at Graving Dock once more.

After six months a compromise was worked out. During the interval the Lough Swilly had been reguaged to 3 ft and the Harbour tramway relaid as mixed gauge. From 1 July 1885, narrow gauge trains were hauled to Middle Quay by a 5ft 3in gauge loco belonging to the Harbour Commissioners. When this was out of action, the Commissioners had to hire one of the redundant 5ft 3in gauge engines of the Lough Swilly at a fee which eroded their increased rent. Before long, the two bodies fell out again. Lough Swilly trains never ran beyond the Graving Dock after 31 December 1887, though wagons could be worked to the ships and to the Great Northern.

The Letterkenny Railway

The town of Letterkenny, which today has a population of 12,000, lies near the head of Lough Swilly. A 5ft 3in gauge railway was authorised by the Letterkenny Railway Act, 1860 but nothing happened and it was 23 years before Letterkenny saw its first train. During this interval, a further five Acts of Parliament were obtained, the route was changed fundamentally and the gauge narrowed.

The original plan was to run eastward from Letterkenny to join the Londonderry & Enniskillen Railway (later the Great Northern) at Cuttymanhill, south of St Johnston. Under the 1863 Act, only the first two miles were left unchanged as far as Pluck. From here the route turned north to join the abandoned Farland Point branch of the Lough Swilly Railway.

Work began but ground to a halt when the money ran out. Extensions of time and capital reorganisations were secured in Acts of 1866, 1871 and 1876. As soon as the last of these had been passed, the Company sought tenders to complete the work. The only result from this exercise was confirmation that the task was impossible within the available capital.

Then came an offer from Messrs McCrea & McFarland to complete the line more cheaply if the gauge were reduced to 3ft. Basil McCrea (1832 – 1907) and John McFarland (1848 – 1926) were both engineering and haulage contractors. They ran the cartage business for the Belfast & Northern Counties Railway in Londonderry. They were to play a long and dominant role in the Lough Swilly Railway. McCrea was a director right up to his death. John McFarland resigned as Chairman in 1917 but his son became a director in 1921 and was Chairman during the last years of railway operation up to 1953.

The response of the Letterkenny Railway, to McCrea and McFarland, was to promote a Bill

Graving Dock Station in 1955, after closure.
(Peter Sunderland)

authorising 3ft gauge. This was defeated in 1878 on the grounds of non – compatability with the Lough Swilly. It was reintroduced and passed in June 1880, this time with powers to regauge the Londonderry & Lough Swilly Railway as well

Work resumed in 1881, both on building the Letterkenny Railway and on restoring the abandoned line across the Trady Embankment. On 30 June 1883, the line opened from Letterkenny to Tooban Junction which, for the next 22 months, offered interchange between the two gauges. The Letterkenny Railway was operated by the Lough Swilly Company.

The line from Londonderry to Buncrana was regauged in one week between 28 March and 4 April 1885. This gave the Lough Swilly a 30½ mile narrow gauge system, albeit that more than half was only worked on behalf of the Letterkenny Company. From June 1887, the owner of the latter became the Irish Board of Works who took possession as mortgagees when the Letterkenny defaulted on loan interest. The Board renewed the working agreement but declined to sell on terms offered by the Lough Swilly.

Tooban Junction
Known originally as Letterkenny Junction, then The Junction, the name Tooban Junction came into use about 1920. In 1883-5, the island platform had a 3ft gauge track on one side and 5ft 3in on the other. Then it became solely a 3ft gauge station with the down line on one side, the up line opposite and a scissors crossing just beyond at the divergence of the Buncrana and Letterkenny routes. There was no road access. The station only functioned for interchange.

From Tooban Junction, the Letterkenny line curved to the south onto the Trady Embankment. Maximum gradient was 1 in 50. The summit was reached near Sallybrook Station whose main customer was the Lagan Creamery.

Expansion Plans
Under the influence of McCrea and McFarland, the Londonderry & Lough Swilly Railway prospered. The value of its shares rallied from next to nothing almost to par value whilst the dividend steadily increased to 7%, a level which was maintained from 1898 until after the First World War. This seemingly incredible state of affairs was achieved by exercising the tightest control on expenditure. The Company had no plans to risk its own money on an expansion into even less inviting territory. From the 1880s, other possibilities began to emerge.

In the County Donegal part of the book, we consider the Acts of 1883, 1889 and 1896, which progressively made it easier for public funds to finance railway expansion into remote areas. A parallel, but non railway, piece of legislation was the Congested Districts Act, 1891. Much of the West of Ireland was a "congested district", a strange term which did not imply a high density of population. What it meant was that there were too many people to be sustained by the prevailing level of economic activity.

No3 arriving at Tooban Junction, from Londonderry in June 1952. (*R H Fullaghar, N E Stead collection*)

Railcar 20 at Stranorlar on a Strabane to Killybegs working. After closure, the two newest railcars were sold to the Isle of Man Railway. They are still there in a semi derelict state having not been used since the 1960s. (John Carter)

Rush hour at Donegal. Railcar 19 has arrived from Killybegs and will continue to Strabane. A railcar going the other way is transferring sacks of mail onto No 16 for Ballyshannon. There is a steam loco at the far end of the Killybegs platform at the head of some wagons (John Carter)

Strabane 1959. Railcar 19 is ready to leave for Stranorlar. The running in board, inviting passengers to change for destinations in Donegal is now displayed at the Ulster Folk & Transport Museum at Cultra. *(Peter Sunderland)*

5 "Meenglas" crossing the River Mourne to enter Strabane with a goods from Stranorlar in 1959. *(Peter Sunderland)*

The Congested Districts Board was able to make grants to assist economic development. One project, which they supported, involved improvements to the harbour at Burtonport on the remote West Coast of Donegal, the better to land fish. The next logical development was the building of a railway, without which the fish still had little hope of reaching consumers in the centres of population.

Under the Light Railways Act, 1896, the Irish Board of Works was empowered to build railways, which would be operated by existing companies. Between 1901 and 1903, 68 route miles were added to the Lough Swilly network, at Government expense.

The distance from Letterkenny to Burtonport is only 28 miles as the crow flies but a straight line would have faced a mountain barrier. The route chosen involved 49¾ miles. It was the shortest practical route to Burtonport which was regarded as the dominant traffic objective. The line missed a number of intermediate villages which had to be content with stations up to four miles away.

The route contained long stretches at the ruling gradient of 1 in 50, including nearly 4 miles from Dunfanaghy Road up to the 474ft summit. This was followed by a 3 mile descent, mostly at the same gradient, to Falcarragh. The greatest engineering work was the 380 yard Owencarrow Viaduct comprising (from the south) 15 girder spans, a short stone embankment and then two masonry arches. The eighth ninth and tenth spans were by far the longest at over 46 yards each. The structure was on a falling gradient of 1 in 50 as far as the tenth span. The eleventh was level, then the remainder rose at 1 in 50. The viaduct was the scene of a fatal accident on 30 January 1925 when high winds caused a train to derail. A six wheel carriage ended upside down on the stone embankment killing four occupants.

The Burtonport line opened throughout on 9 March 1903 with two trains each way on weekdays and one on Sundays. A third weekday train was soon added at the insistence of the Board of Works,

who had paid for the line to be built. They weren't paying for its operation and the Lough Swilly Company begrudged the extra train, claiming that patronage was minimal. In 1916, they were prevented from withdrawing it by an injunction. Journey time was typically 4 hours 10 minutes for the 74½ miles from Londonderry.

Carndonagh

The other Government extension, the 18¼ mile Buncrana & Carndonagh Light Railway had opened on 1 July 1901. First promoted in 1884, the scheme had failed to qualify for Government assistance under the 1883 Act, not least because the Board of Works did not believe the cost estimated by the promoters, who were told to prepare their plans more thoroughly.

The new line climbed steeply for seven miles out of Buncrana. Maximum gradient was 1in 50 and the 332ft summit was reached a mile beyond Drumfries. The descent was equally severe on the four mile descent to Clonmany, the only passing loop and block post. There was renewed climbing to Ballyliffin, then another steep drop almost to Sea level at Rashenny, the most northerly station in Ireland. From here the line ran close to Trawbegga Bay. Carndoagh Halt was opened about 1930. The last 1½ miles were at a slight incline leading to the terminus on the north side of Carndonagh, a town with less than 1,000 population in the 1920s though it is a lot more now.

Dissention

The working of the extensions was the subject of continual argument between the Irish Board of Works and the Railway Company. The former complained that the lines were not being operated properly. The latter accused the Board of not having built them satisfactorily.

In 1905, the Board of Works commissioned an inquiry from Joseph Tatlow, Manager of the Midland

Great Western Railway. This resolved nothing until 1909 when the two sides agreed to accept arbitration by Sir Charles Scotter, Chairman of the London & South Western Railway. The "Scotter Award" comprised eight items, one of which was a grant from the Board of Works allowing the Railway to purchase two locomotives, a carriage, two fish vans and 27 wagons.

The acrimony continued and Tatlow was asked to report again in 1917. He found a great deal wrong with the undertaking, which was placed in receivership under the terms of the Light Railways Act, 1896. John McFadden resigned as Chairman having been caught using Company resources to maintain his private estate. Henry Hunt became Receiver & Manager with a budget from the Board of Works to put things right.

This process began in the midst of the First World War, which was followed by the Irish Troubles and imposition of the Border. When "normality" began to return after 1923, it was a very different World in which it scarcely mattered whether there were two trains a day to Burtonport, the Company's wish, or three as the Board of Works had always tried to insist.

Motive Power

In standard gauge days, the Lough Swilly Railway had employed a total of six locomotives. The last two, Nos 4 "St Patrick" and 5 "St Columb" were 0 – 6 – 0 tanks, built by Sharp, Stewart of Manchester in 1876 and 1879. After gauge conversion in 1885, they were sold to the Cork, Bandon & South Coast Railway.

22 narrow gauge locos were acquired between 1882 and 1912. Of particular interest were the heavy engines built after 1899 for the extensions. Weighing 40 tons, Nos 5 and 6 (later 15 and 16) were built for the Carndonagh line. The Board of Works complained, and even brought legal proceedings when they were used elsewhere. The Andrew Barclay 4 – 6 – 0Ts, built in 1902 for the Burtonport extension, were smaller at 35 tons. Designed to the specification of the Board of Works, the Railway considered them inadequate and in 1905 acquired two 4 – 8 – 0 tender engines from Hudswell Clarke of Leeds. Weighing 37 tons plus 21 for the tender, they could get from Letterkenny to Burtonport and back without recoaling.

Nos 13 and 14 were 4 – 6 – 2Ts authorised in 1910 under the "Scotter Award". They weighed 41½ tons but were eclipsed by the new Nos 5 and 6 of 1912 which were the only 4 – 8 – 4 tanks in the British Isles. Weighing 51 tons, they were by far the largest engines ever used on the Irish narrow gauge.

The Long Decline

After 1924, the Railway received a subsidy from the Irish Free State. From the following year, a contribution was also received from Northern Ireland. In 1927, the Company asked the neighbouring County Donegal Joint Committee if it was interested in taking it over. Nothing came of this, nor of quotations received from Armstrong Whitworth for two diesel railcars.

Instead, in 1931, the Company agreed with both Governments that it would acquire competing road services with a view to eventual closure of the Railway.

The Carndonagh extension closed completely on 30 November 1935. At the same time, passenger trains to Buncrana were limited to Thursdays and Saturdays in Summer only. The displaced traffic was taken over by the Company's own buses and lorries. These were not affected by the 1935 setting up of the Northern Ireland Road Transport Board because their mileage in Northern Ireland was minimal.

The Burtonport Extension closed on 3 June 1940, when passenger trains were also withdrawn between Londonderry and Letterkenny. Very soon, the fuel shortage of the Second World War began to bite. Track lifting had begun from Burtonport in August 1940 and had quickly reached Gweedore. There it stopped and on 3 February 1941 the truncated line reopened for goods. From March 1943, passengers were also carried on the daily Londonderry to Gweedore "mixed".

For the next six years, the Railway did good business, carrying more passengers than it had pre 1930. But it was a short lived respite with minimal maintenance and even periods when the locomotives had to burn turf instead of coal.

Regular traffic ceased again between Letterkenny and Gweedore in January 1947, though a few special consignments were carried until June that year. After that, goods traffic continued between Londonderry and Letterkenny, the train conveying a passenger coach as brake van so a few tickets were still sold.

Passenger trains to Buncrana ceased in September 1948 apart from Bank Holiday specials which ran until August 1951. It only remained for the lorry fleet to be built up sufficient for the Company to apply to the two Governments for permission to close completely. No 10 hauled the last train on Saturday 8 August 1953, 14 loaded cattle wagons from Letterkenny.

Signalling

Early single line security was by a staff. There were three sections: Londonderry to Tooban, Tooban to Buncrana and Tooban to Letterkenny. If a second train wanted to follow in the same direction, the arrangement seems to have been haphazard. The staff was carried by whichever train required the key to work intermediate sidings.

On Sunday 21 June 1891, there was a head on collision 1¼ miles outside Londonderry. A heavily loaded troop train was travelling towards Tooban Junction behind No 5. It was hit by No 2 at the head of an empty train, which should not have been in the section. The driver and firemen on No 2 were killed, those on No 5 survived by jumping clear before impact. 14 troops were injured. It emerged in

4 – 4 – 0 No 74 pauses at Bundoran Junction with a goods from Omagh to Enniskillen in September 1957.

(John Carter)

3T has arrived at Arigna in August 1958. The kids are about to have a bath, just as soon as Mother can fill the tub with hot water from the loco. *(John Carter)*

Enniskillen from the south east with the SL&NC railbus just visible in the bay platform and one of their 0 – 6 – 4 tanks between the GN locos in the shed yard on the right.   *(John Carter)*

0 – 6 – 0 No 179 pauses at Crumlin with a local from Belfast Great Victoria St to Antrim in August 1958.

(J C W Halliday)

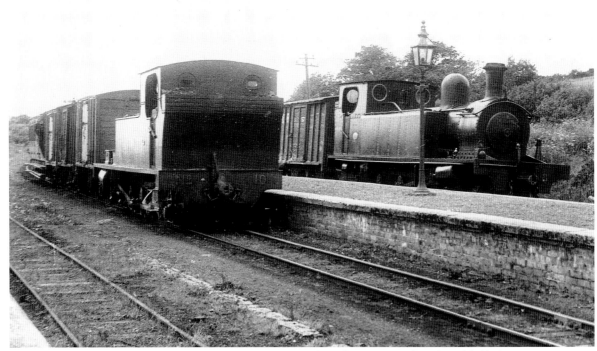

Tooban Junction on 27 June 1952 with 4 – 6 – 2T No 10 arriving from Buncrana, coming alongside No 8 waiting on the up line.   *(R H Fullaghar/ N E Stead collection)*

4 – 8 – 0 No 12, waiting to leave Burtonport on 24 June 1937.   *(H C Casserley)*

evidence that the signalman at Tooban Junction was illiterate and that his practice had been to get a passenger, neighbour or anyone available to read instructions to him.

This accident, which was aggravated by the absence of continuous brakes, came only two years after the disaster near Armagh on the Great Northern. During 1892, the Lough Swilly complied with new regulations by fitting passenger trains with automatic vacuum brakes. It also introduced staff and ticket working, reinforced with block telegraph.

The Carndonagh extension was equipped from the start with electric tokens, the line being split into two sections, with a passing loop at Clonmany. The Board of Works also financed the electric token system on the Letterkenny to Burtonport extension with passing places at Creeslough and Gweedore. Tooban to Letterkenny was converted to electric token working in 1909 but Londonderry to Buncrana always retained staff and ticket. Fahan became an additional block post about 1918.

Such were the systems installed. The second Tatlow inquiry of 1917 found most of them not working. The electric token is entirely dependent on maintaining the block cable throughout the length of the line.

The City of Derry Tramways Company

Despite its title, this venture was never an attempt to establish a network of street tramways around the City but just one short line associated with the Lough Swilly Railway.

To overcome the inconvenience of Graving Dock Station, the City of Derry Tramways Company was incorporated in 1893 to lay a 4ft 8½ in gauge horse tramway between the Lough Swilly and Great Northern Stations. It was financed, substantially, by Messrs McCrea and McFarland. Nine open top double deck tramcars were ordered. The tramway opened during 1897 and lasted until February 1919. It ran parallel to the Harbour Commissioners' mixed gauge goods line, affording the unique spectacle of three different gauges over this 1½ mile section.

The Lough Swilly Steamboat Company

The Pier at Fahan was in Railway ownership from its opening in 1868. The steamers, however, remained in various private hands until 1877 when Messrs McCrea & McFarland formed the Lough Swilly Steamboat Company. This body ran up to three small paddle steamers linking Fahan with piers on the West side of the Lough.

From 1905, the Company's' "flagship" was the 140 ton paddle steamer "Lake of Shadows". The winter timetable for 1913 shows three sailings each weekday to Rathmullan, connecting out of trains from Londonderry. There was a sailing once a week to Portsalon on a Tuesday. The Ramelton service ran every weekday but a fixed timetable was impossible. Ramelton Pier, on the River Leannan, could only be used for three hours on either side of high tide.

On 7 January 1920, Fahan Pier was destroyed in a gale. It took until the end of 1922 for the Railway Company to build a replacement. The Railway itself took over the steamers in 1923. "Lake of Shadows" was withdrawn in 1929 in favour of smaller motor boats. The service lasted in Railway ownership until 1952.

Horse trams passing near Middle Quay. The property to the right of the further tram, with posters in the window, is a booking office for cross channel rail and steamer tickets. *(Martin Bairstow collection)*

4 – 6 – 2T No 15 at Newtown with the 2.15pm goods from Letterkenny to Londonderry on 20 April 1953. The passenger brake catered for enthusiasts who still wished to travel.
(H C Casserley)

The same train crossing the Trady Embankment
(H C Casserley)

Opposite Upper No 4 has reached the summit at Hill of Howth during the last summer of operation in 1958.
(J C W Halliday)

Lower "Dick" pulls away from Fintona on another trip to the Junction. Has the lady missed it and is she now trying to catch it up? *(E S Russell/ Colour Rail)*

Newtown, looking towards Londonderry in April 1953. No 5 has left its train in the Platform, whilst it shunts the Yard.
(H C Casserley)

The Lough Swilly Ferry

Sailings resumed on 14 May 2004, with the inauguration of a Summer only car ferry between Buncrana and Rathmullan. Operated by the Lough Foyle Ferry Company, the "Foyle Rambler" makes the crossing hourly.

Arranmore Island Ferry

Another former Lough Swilly Railway outpost to have gained a ferry is Burtonport. A century ago, the Congested Districts Board supported a railway to Burtonport in the hope that it would generate economic development. Today, the Gaeltacht (Irish Language) Authority supports a passenger and car ferry to Arranmore for much the same reason.

Commencing in 1992, the route has acquired four of the small "Island" class small car ferries built for Caledonian MacBrayne in 1972 – 74. "Kilbrannan" was renamed "Arainn Mhor" but "Coll", "Morven" and "Rhum" retain their Scottish names. The 25 minute crossing operates all year round, with up to eight services a day.

The 120 ton paddle steamer "Innishowen" served the Lough Swilly Company from 1881 until 1912. Pictured at Ramelton, it offered Spartan accommodation in two classes.
(Martin Bairstow collection)

"Coll" leaving Burtonport for Arranmore in September 2003.
(Martin Bairstow)

Still trading as the Londonderry & Lough Swilly Railway Company, a line up of buses outside Carndonagh Station in September 2003.
(Martin Bairstow)

4 – 6 – 2T No 10 at Fahan with a train for Buncrana about 1950. The engine was built by Kerr Stuart in 1904. The station is now the Railway Tavern and Firebox Grill.

A Letterkenny to Londonderry goods behind 4 – 6 – 2T No 8 at Newton Cunningham in 1952. The passenger brake serving as guards van is available for parcels and even the occasional fare paying passenger.
(R H Fullaghar, N E Stead collection)

4 - 6 – 0T No 2 awaiting the attention of the scrapman at Pennyburn in March 1954. Nos 1 to 4 were built by Andrew Barclay, Kilmarnock in 1902. Pennyburn loco shed and workshops were just north of Graving Dock Station. The scrapman (or liquidator) finally caught up with the Londonderry & Lough Swilly Railway Company in April 2014 when it ceased bus operations in response to a winding up petition from H M Revenue & Customs. *(John Oxley)*

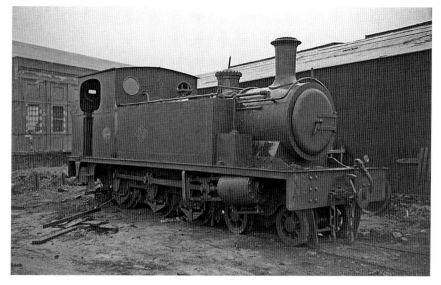

The County Donegal Railway

No 4 "Meenglas" with the Letterkenny goods, crossing the Foyle into Lifford in August 1956. During January 1960, the bridge was converted into a road for use by Joint Committee buses and lorries. It was demolished following opening of the present road bridge in 1964. *(Peter Sunderland)*

I was still at school, in the 1960s, when I first read E M Patterson's standard work on the County Donegal. I realised then that this had been a remarkable railway. It was the largest narrow gauge system in the British Isles, a pioneer with diesel railcars. It was half owned by British Railways though only one station lay within the United Kingdom. Perhaps the most remarkable feature was that some of the network existed at all and lasted as long as it did in such a remote and sparsely populated region.

In 1841 the population of County Donegal was 296,000. That was before the Railway Age but, more important, it was before the Potato Famine, which caused so many people to choose between emigration and starvation. The decline in population continued well into the twentieth century. By the 1950s, the last decade of the Railway, the figure was down to 130,000.

What the County lacks in population and industry, it makes up in scenery. By the time David Beeken and I first visited Donegal Town in 1999, the place was certainly open for tourists. Our difficulty was parking the car, which we had hired from Sligo, the nearest we could get by train. Back in the days of the Railway, it was a long trek from mainland Britain and people didn't have the time or money except for the few who made it their annual holiday.

Tourists from the Continent or further afield would have been virtually unknown. Philippa, my wife, visited County Donegal in 1960 and stayed in accommodation without mains electricity. It was a very different place we found 39 years later.

The Finn Valley Railway

The oldest section of what became the County Donegal Railway pre-dated the 3ft gauge. The 13½ miles between Strabane and Stranorlar were built to the Irish gauge of 5ft 3in.

The Finn Valley Railway was incorporated by an Act of May 1860. Foremost amongst the promoters was Viscount Lifford who tried to persuade other landowners to invest in the project with hopelessly optimistic forecasts of likely dividends. It was a familiar story. The project cost more than expected and took longer to build. Shareholders were slow at paying calls and the cheapest contractor turned out not to have been the best.

At Strabane, the new line joined the existing Irish North Western Railway some 660 yards south of the station at what became Finn Valley Junction. A rent had to be paid for use of this short stretch of track. In addition, it was agreed that the Finn Valley would give the Irish North Western 35% of gross receipts for working its line. This was a common practice

No 12 with a Strabane train at Stranorlar in August 1956.
(Peter Sunderland)

whereby small companies avoided having to obtain their own motive power, rolling stock and crews.

Opening took place on 7 September 1863 with three trains each way between Strabane and Stranorlar, weekdays only. Over the following years, the Finn Valley blamed its relative poverty on the arrangements with the Irish North Western. In 1871 the Finn Valley promoted a Bill for an independent entry to Strabane involving a new bridge over the River Mourne and a separate station nearer to the centre of Strabane. The Bill was defeated in the House of Lords. From 1 November 1872, the Company achieved a small measure of independence, purchasing its own carriages but still hiring in Irish North Western locos and footplate crews.

The West Donegal Railway

Some Finn Valley shareholders, including the Chairman, saw the salvation of their Railway in an extension to Donegal Town but they couldn't persuade sufficient people there to come forward with money.

By the mid 1870s the 3ft gauge had been tried successfully in County Antrim and this offered a cheaper alternative. In July 1879 the West Donegal Railway was incorporated under the Chairmanship of Viscount Lifford with powers to build a 3ft gauge line covering the 18 miles from Stranorlar to Donegal.

They accepted the lowest tender, knocked it down a bit, then persuaded the contractor to accept part payment in shares. Despite all this, they couldn't raise enough capital to go all the way to Donegal. When the line opened on 25 April 1882, it terminated four miles short at Druminnin, a remote station which was later called Lough Eske. Three trains were provided each way daily except Sundays. Responsibility for operation was assumed by the Finn Valley Railway who, though unable to

provide motive power for their own line, managed to purchase three 2-4-0 tank engines from Sharp Stewart of Manchester. These were named "Alice", "Blanche" and "Lydia" after relatives of Viscount Lifford.

Between Stranorlar and Druminnin, the Railway encountered the bleakest and most difficult terrain of the eventual 124½ mile system. The first 5 miles involved a continuous climb at between 1 in 50 and 1 in 60. Then the line was virtually flat as it traversed the moorland plateau close to Lough Mourne. A short climb at 1 in 97 led to the 592 feet summit at Derg Bridge Halt, the highest point reached by the County Donegal. The descent began at 1 in 64, steepening slightly to 1 in 60 as the Railway assumed its position on a ledge cut along the steep south grade of the Barnesmoor Gap.

By contrast with all this, the unbuilt four miles to Donegal passed through easy country. In 1886 a fresh effort was made to raise capital and the West Donegal finally reached the County Town on 16 September 1889.

Government Help – Killybegs and Glenties

Viscount Lifford did not live to see his Railway reach Donegal. He died in 1888. Hitherto, railway building in remote areas had depended on people like him, the only ones with any money. Some landowners did accept that responsibility went with privilege and tried to get public works moving. In the most difficult areas, this was not sufficient. Under the Light Railways (Ireland) Act 1889, the possibility of state aid was introduced. During the next two years, Light Railway Orders were obtained for two extensions: 19 miles from Donegal to Killybegs and 24 miles from Stranorlar to Glenties. Both lines were to be financed substantially by Government loans.

The Killybegs route was something of a rollercoaster with several stretches at 1 in 40. At various points the line was close to the Sea but, each time a peninsula jutted out, the line had to climb over the corresponding ridge of high land projecting inland. The final approach to Killybegs was along the Sea wall. The terminus was close to the Harbour. A siding extended on to the pier. The line opened on 18 August 1893.

Glenties followed on 3 June 1895. This served a very sparsely populated area, passing through the upper Finn Valley until it reached a summit just before Fintown. After a level stretch the length of Lough Finn, it descended through the Shallogan Valley to the small town of Glenties.

The Donegal Railway – Regauging the Finn Valley
The Killybegs Order had been granted to the West Donegal Company, that for Glenties to the Finn Valley. In June 1892, the two merged to form the Donegal Railway. This Company obtained powers in 1893 to reduce the gauge of the original Finn Valley section to 3ft and build a new approach to Strabane with its own station adjoining that of the Great Northern Railway. Preparatory work was completed so that the final gauge conversion would disrupt traffic only on a single weekend. Beginning on Friday evening 13 July 1894, gangs toiled the whole way from Strabane to Stranorlar, moving one rail in and re-spiking it to the sleepers. Narrow gauge traffic was running on Monday morning, 16 July.

Strabane to Londonderry
Removal of one break of gauge had simply created another. A good proportion of traffic from the Donegal was heading for the City or Port of Londonderry but goods had to be transhipped manually at Strabane. In 1896, despite opposition from the Great Northern, the Donegal gained powers for its own line which would run on the east side of the River Foyle to a terminus called Victoria Road, adjacent to the Carlisle Bridge in Londonderry. This project was a commercial venture by the Donegal not qualifying for any Government assistance. It opened to goods on 1 August 1900 and to passengers five days later.

The Londonderry Port & Harbour Commissioners extended their mixed gauge quayside line so as to connect the Donegal with the Belfast & Northern Counties Station at Waterside. This meant also that wagons could reach Londonderry's other two stations on the west side of the Foyle because they could be winched across the lower deck of the Carlisle Bridge.

Ballyshannon
A 15¾ mile branch from Donegal to Ballyshannon was authorised by the same 1896 Act but capital was rather slow coming in and no work had been done by the time the powers expired after five years. A new Act was obtained in 1902 and the line opened on 21 September 1905.

2 – 6 – 4T No 18 "Killybegs" backing onto its train at Londonderry Victoria Road on 7 May 1920.
(LCGB Ken Nunn Collection)

It left Donegal Station in an easterly direction quickly diverging from the line to Stranorlar. The maximum gradient was 1 in 60. Intermediate population was minimal. The terminus was on the north side of Ballyshannon, about a mile from the Great Northern Station.

The County Donegal Railways Joint Committee
In 1903, the Midland Railway (of England) absorbed the Belfast & Northern Counties Railway, henceforth managed through its Northern Counties Committee. This brought the Midland to Londonderry Waterside. The Company's next move was to make an offer for the Donegal Railway. This alarmed the Great Northern whose opposition might have frustrated the necessary enabling legislation. Compromise emerged and the Donegal was taken over by the two Companies jointly. Because the Great Northern had its own parallel route between Strabane and Londonderry, the Midland insisted on taking sole ownership of that section. For operating purposes it came under the County Donegal Railways Joint Committee which began to function on 1 May 1906.

The Strabane & Letterkenny Railway
The small town of Convoy was home to a woollen industry. In 1903 the Strabane, Raphoe & Convoy Railway had been incorporated to build a nine mile line between these places. The following year, a further Act extended the proposed line by ten miles to Letterkenny. The project was financed substantially by the Joint Committee who operated the nominally independent Strabane & Letterkenny from its opening on 1 January 1909. The route included gradients up to 1 in 45. Just before the

terminus, the line crossed over the Londonderry & Lough Swilly Railway to end in its own adjacent station where the tracks of the two Railways joined.

This brought the County Donegal network to its maximum of 124½ route miles which included 14½ owned by the Midland and 19¼ by the Strabane & Letterkenny.

Train Services and Motive Power
The standard passenger service involved three trains each way over the more remote sections. In 1906 there were six trains between Londonderry and Strabane, five between there and Stranorlar and four thence to Donegal. This was broadly the pattern until closure. The Londonderry route declined, particularly after 1934, when the loco shed closed and the service fell to three each way. By contrast, Letterkenny started with three trains a day and ended up in railcar days with five.

Sunday trains mostly ran in Summer only, apart from a morning working from Ballyshannon to Rossnowlagh and back which ran for the benefit of Church goers, rather like Kirk Braddan in the Isle of Man. So that railcars could work this and any excursions terminating at Rossnowlagh, the turntable from Londonderry Shed was re-installed there.

To work all this traffic, the Joint Committee owned a total of 23 steam locomotives, though never all at once. Two of the original 2-4-0 tanks were scrapped in 1912 to make way for the final acquisitions. Eventually, in 1928, the newest three engines adopted the names of the original three. Further renaming and renumbering took place in 1937.

Great Northern and Lough Swilly buses in front of Letterkenny (Co Donegal) station in 1958. The building s still in use today as a booking office for Bus Eireann. *(J C W Halliday)*

Original Number	Original Name	Type	Built	Withdrawn	New Number	New Name
1	Alice	2-4-0T	1881	1926		
2	Blanche	2-4-0T	1881	1912		
3	Lydia	2-4-0T	1881	1912		
4	Meenglas	4-6-0T	1893	1935		
5	Drumboe	4-6-0T	1893	1931		
6	Inver	4-6-0T	1893	1931		
7	Finn	4-6-0T	1893	1931		
8	Foyle	4-6-0T	1893	1937		
9	Columbkille	4-6-0T	1893	1937		
10	Sir James	4-4-4T	1902	1933		
11	Hercules	4-4-4T	1902	1933		
12	Eske	4-6-4T	1904	1954	9	Eske
13	Owenea	4-6-4T	1904	1949	10	Owenea
14	Erne	4-6-4T	1904	1960	11	Erne
15	Mourne	4-6-4T	1904	1940	12	Mourne
16	Donegal	2-6-4T	1907	1960	4	Meenglas
17	Glenties	2-6-4T	1907	1960	5	Drumboe
18	Killybegs	2-6-4T	1907	1960	6	Columbkille
19	Letterkenny	2-6-4T	1908	1940	7	Finn
20	Raphoe	2-6-4T	1908	1955	8	Foyle
21	Ballyshannon	2-6-4T	1912	1960	1	Alice
2A	Strabane	2-6-4T	1912	1960	2	Blanche
3A	Stranorlar	2-6-4T	1912	1960	3	Lydia

1 to 3 were built by Sharp, Stewart of Manchester, 4 to 11 by Nielson of Glasgow and 12 onwards by Nasmyth Wilson of Patricroft. The fleet was distributed amongst seven loco sheds. The five termini each had a shed in addition to Strabane and Stranorlar.

Railcars

The original petrol driven railcar was no more than an inspection vehicle. Built in 1907 and now in the Ulster Transport Museum at Cultra, it was little used after 1920 until a shortage of coal in 1926 caused it to be run in service on the Glenties line. Seating six pasengers, it also carried the mail and tempted the County Donegal to look for something similar but larger. The result was the purchase and regauging of two vehicles made redundant on closure to passengers of the Derwent Valley Railway, near York (see *Railways in East Yorkshire Volume One*). Seating 17 passengers each, they lasted until 1934.

By that time, the Joint Committee had experimented with further petrol railcars but the breakthrough came in 1931 with Nos 7 and 8. These were the first diesel railcars to operate in the British Isles. Seating 32, they were carried on six wheels, four supporting the body towards the rear and two projecting out in front of the long nose which housed the engine. Nos 7 and 8 lasted until 1949 spending much of their time on the Ballyshannon branch.

No 12 of 1934 was the first of the Gardner – Walker railcars mounted on two bogies. Eventually nine of these were owned culminating in the streamlined Nos 19 and 20 of 1950/51. Unfortunately, not all nine existed at the same time because No 17, of 1938 was destroyed in a collision

on 29 August 1949 killing the driver and two passengers. The driver had just left Donegal for Ballyshannon without the single line staff and in blissful ignorance of a special freight coming the other way behind 4-6-4T No 10 "Owenea".

By the mid 1930s, railcars had taken over most passenger working except between Strabane and Londonderry which always remained steam operated. Already a number of halts had been opened earlier in the century to supplement the original stations. The practice grew that railcars would stop anywhere to pick up or deposit passengers at the lineside. In 1944 the management decided that this was impractical and published a list of authorised stopping places. Many of these were at level crossings and took their names from the resident crossing keepers.

The railcars were unidirectional so turntables had to be installed at termini which didn't already have one. They could pull a modest tail load.

Signalling

The single line was controlled by electric train staff. Besides the junctions and termini, there were passing loops and block posts at Donemana, Castlefinn, Lough Eske, Inver, Fintown, Ballintra, Raphoe and Glenmaquin. Some of these were dispensed with in later years.

All the block posts had signal boxes or ground frames. There were East and West Cabins at Stranorlar. Signals were not interlocked with staff instruments. At intermediate stations, points were released by a key attached to the single line staff.

There were 66 level crossings at which public roads crossed the railway. Apart from those close to

No 12 at Killybegs in 1958. This was then the most westerly station on British Railways, joint with the Great Northern Railway (Ireland). *(J C W Halliday)*

No 19 leaving Stranorlar for Donegal in August 1956. *(Peter Sunderland)*

This page top No 4 "Meenglas" being hand coaled at Strabane in 1959. *(Peter Sunderland)*

Middle Railcar 16 leaving Lifford for Strabane in August 1956. The bonnet flaps were often left open to help prevent the engine from overheating. *(Peter Sunderland)*

Bottom No 11 "Phoenix" was a 0 – 4 – 0 diesel loco rebuilt in 1932 out of an unsuccessful "steam tractor", with which the Clogher Valley Railway had found itself saddled. It was used mainly at Strabane, where it is seen in August 1959 shunting coach No 59, one of the corridor vehicles built in 1928 for the Ballymena to Larne "boat trains". *(Peter Sunderland)*

In 1958, Mike Swift travelled in the brake third attached to the 12.50pm goods from Letterkenny. The object was to change at Strabane on to the 3pm goods to Stranorlar. But the train was delayed by Customs at Lifford. The guard advised Mike and his friend to walk the last mile to Strabane but the driver uncoupled "Meenglas" from the train and took them light engine. At Strabane they joined the passenger brake attached to the Stranorlar goods hauled by "Drumboe". They were not alone, there were other passengers travelling on the "goods".

During the Second World War, military personnel and equipment were not allowed to use the GN between Londonderry and Strabane because the line passed through neutral territory. Individuals might travel dressed in "civvies" but heavy stuff had a long detour via Cookstown. Consideration may have been given to re-gauging the alternative route out of Londonderry Victoria Road but the idea was not pursued.

a station, they were manned by resident crossing keepers. Very few had a telephone or block bell. Gates were opened in response to train whistles, the sight of smoke or knowledge of the timetable. A few of the crossings were protected by signals. Otherwise, discs and lamps on the gates themselves were their only defence.

Decline and Closure

The Railway managed quite well during the Second World War. Whilst running mainly through the neutral Free State, it had access to fuel in Northern Ireland.

Traffic on the Glenties branch was minimal and the passenger service was withdrawn in December 1947. Spasmodic freight continued until March 1952.

On 1 April 1949, the lines making up the Northern Counties Committee passed from British Railways to the Ulster Transport Authority. This body closed the Strabane to Londonderry section on the last day of 1954.

The remainder of the County Donegal system continued to function through the 1950s. It even innovated with the introduction of a container service between Dublin and Letterkenny. The containers were Customs sealed for the journey through Northern Ireland and were transhipped at Strabane onto adapted narrow gauge wagons, far easier than the traditional method of freight handling.

But continued operation was at the expense of skimping on maintenance and renewals. In 1957/58,

one owning company, the Great Northern Railway was all but destroyed. The other joint owner was British Railways. Neither was interested in investing in the Donegal. The Railway was just waiting for the roads to be brought up to standard before it could be closed.

In 1957, the Joint Committee asked the Irish Transport Tribunal for permission to close the Ballyshannon branch and replace it with buses and lorries. Consent was conditional upon road improvements. Two years later, Ballyshannon was still running when the Joint Committee sought and received permission to close the entire remaining network. The last trains were scheduled for 31 December 1959. Railcar No 12 worked the 4.05pm from Killybegs. On arrival at Stranorlar, it was overwhelmed by the traffic on offer so it gave way to 2-6-4T No 5 'Drumboe' with five carriages for the final run into Strabane.

Replacement buses and lorries were unable to cross the weak road bridge over the Foyle into Strabane so the Joint Committee quickly converted the Letterkenny line bridge for road use. Pending this, goods trains continued between Strabane and Stranorlar for the first five weeks of 1960. Then the entire railway was demolished. Belfast Transport Museum acquired loco No 2 'Blanche', the diesel shunter 'Phoenix', railcar No 10 and a trailer. These are now at Cultra.

The rest of the rolling stock was auctioned off on 1 March 1961. The two newest railcars went to the Isle of Man Railway. Four steam locos, three railcars and numerous carriages and wagons were

No12 at Killybegs in August 1958, on the turntable made out of the frame of a 2 – 6 –4T engine.

(Mike Swift)

Most ordinary trains had been railcar operated since the 1930s but steam still appeared on specials. No 2 "Blanche" passing Killygordon with an 11 coach bank holiday excursion from Strabane to Ballyshannon on 3 August 1959. Killygordon Station is now well restored as a private residence. *(J C W Halliday)*

The same train entering the passing loop at Lough Eske which came before the single platform station. The fourth vehicle is one of the Ballymena & Larne "boat train" carriages. *(John Carter)*

The trestle table used by Customs at Clady. If these chaps are the Customs Officers, they look shiftier than the smugglers. *(J C W Halliday)*

Luxury travel in County Donegal. Lighting is by acetylene. Only coaches 57 to 59, from the Ballymena & Larne had electric lights.
(Peter Sunderland)

When the GN was divided between CIE and the UTA in 1958, the half share in the Co Donegal Joint was left in limbo. The GNR Board remained in existence until 1977, just to clear up this matter. The British Railways half share was sold to CIE in 1967. The GN half went to CIE on dissolution of the Joint Committee in 1971. CIE buses became Bus Eireann in 1987.

purchased by Dr Ralph Cox from New Jersey, USA, who intended to ship them there but was never able to do so.

The Joint Committee continued to function until 1971, when its buses and lorries were taken over by CIE. Until closure of the "Derry Road" in February 1965, these had continued to exchange traffic with trains at Strabane station.

Preservation

The County Donegal had its aficionados but most, if not all, accepted the inevitable when it closed. The best which could be salvaged were the relics taken to Belfast Transport Museum and memories brought together in Dr Patterson's history book.

A generation was to pass before people had the time and money to think of preservation, as we now know it. In that time, everything had disappeared, or would have done but for the failed effort of Dr Cox. It is thanks to him that three steam engines, three railcars and other rolling stock were given the chance to survive.

After working the demolition trains, the locos and railcars were just left where they finished up. Eventually, by the early 1970s, CIE wanted them moving and a home was found at Victoria Road Station in Londonderry. In 1978, that establishment changed hands and the stock moved to Shanes Castle, near Antrim. In 1989, some of it moved again

to Foyle Road, Londonderry where a narrow gauge railway was to be built on the old Great Northern trackbed towards Carrigans.

Railcar 18 has been superbly restored and has run on 2½ miles of line which starts at Foyle Road but does not yet reach a destination station. Nor is it likely to following the closure of the enterprise in 2000 and the fall out between Derry City Council and the North West of Ireland Railway Preservation Society. In 2003, No 18 moved to Fintown where 1½ miles of track have been relaid alongside the Lough towards Glenties. Again the line does not yet reach a destination station.

Further south, Donegal Station has been restored as a heritage centre by the County Donegal Railway Restoration Society. It is hoped to rebuild the line for ¾ mile towards Stranorlar as far as Garells Crossing where the train would provide a link between town centre and a shopping complex. No 5 'Drumboe' is being restored to working order along with railcar 15.

The dream of an operational railway seems a long way off and is, perhaps, not helped by the fragmented approach. In October 2001, the North West of Ireland and Donegal Restoration Societies declared an 'iron handshake', a resolve to connect their two projects with a narrow gauge link all the way from Londonderry to Donegal.

Railcar No 18, built 1940, at Foyle Road in May 1995. *(Martin Bairstow)*

Fintown Station in September 2003 with loco No 6 and a former Belgian tram serving as carriage. *(Martin Bairstow)*

Donegal Station has been superbly restored as a heritage centre but will it ever see another train?

(Martin Bairstow)

Interior view of County Donegal Railcar No 18.

(Martin Bairstow

The Cavan & Leitrim Railway

2 – 6 – 0T No 4T taking water at Mohill, whilst working the 7pm Ballinamore to Dromod in July 1957.

(Peter Sunderland)

Unusually for Ireland, this 3ft gauge line survived on coal traffic. It ran from 1887 until 1959, steam operated to the end. It outlived most of the other narrow gauge Irish lines. It gave a lease of life to some of their redundant engines.

The railway came about under the Tramways & Public Companies (Ireland) Act 1883 which allowed the Grand Juries, predecessors of the County Councils, to guarantee dividends. This they did at the expense of ratepayers but they were reimbursed half the cost by Central Government.

The Cavan & Leitrim comprised a main line, 34 miles in length from Dromod to Belturbet with a 15 mile branch from Ballinamore to Arigna. The branch was known as the Tramway. After three miles it came to the roadside and for the next 12 miles it followed every dip and bend in the adjacent rather primitive highway. The steepest gradient was 1 in 25, encountered on the Tramway near Creagh Halt. There were several short stretches exceeding 1 in 40 on the main line.

Passenger traffic was minimal and trains usually comprised a single carriage as part of a 'mixed'. Journey times from Ballinamore were about an hour to Dromod, an hour to Belturbet but an hour and a half to Arigna. For most of the line's existence, the standard service comprised three trains a day on each section, but with no service on Sundays.

The exception to the generally sparse flow of passengers was on 'Fair Days' when the capacity of both the passenger and cattle wagon fleets was stretched. Monaghan Fair was held at Mohill on 25 February each year in honour of St Manachan. Around 1900, there was an annual pilgrimage to Arigna in September. 1 August saw pilgrim traffic to Drumshanbo right up to closure whilst specials were arranged for the Orangemen on 12 July. There were seaside excursions to Bundoran via Belturbet.

Single line security was by staff and ticket. On the main line there were passing loops at Mohill, Ballinamore, Bawnboy Road and Ballyconnell. Sidings at other stations were on Annett's locks. The Ballinamore to Arigna branch was initially worked on the one engine in steam principle. In 1893, staff and ticket was introduced as far as Drumshanbo and from there to Arigna in 1905. The extension (see later) was a separate staff and ticket section.

A kind of time interval system was used at times of extraordinary traffic, such as Fair Days, on the Arigna branch. It took over an hour to traverse the Ballinamore to Drumshanbo section so they despatched consecutive trains without waiting for

4 – 4 – 0 No 8 "Queen Victoria" at Belturbet with the 11.45 to Dromod on 13 May 1920.
    (LCGB – Ken Nunn collection)

Arigna Station in July 1957. The extension to the mines curved sharply to the right, immediately beyond the platform.   (Peter Sunderland)

4 – 6 – 0T No 3T leaving Drumshanbo with the afternoon "mixed" from Ballinamore to Arigna in July 1957.
    (Peter Sunderland)

the previous one to clear. The practice was illegal on passenger lines but was far enough away for the Board of Trade not to know about it. In late Great Southern and CIE days, they continued the practice but with intermediate telephones replacing the time interval. There were 33 level crossings on the main line. Apart from those at stations and a handful of occupational crossings, gates were operated by resident crossing keepers. Some had protecting signals in one direction or the other.

12 coaches were ordered for the opening of the line: four third class and eight composites. This gave far too much first class accommodation so three of the composites were later converted to all third. First class passengers were provided with upholstered arm chairs. Third class sat on longitudinal wooden seats.

No 7 composite was rebuilt in 1953 using two single deck bus bodies. Inside it had bus seats, all third class. No 1 composite was rebuilt as late as 1958, fully enclosed at each end without verandas. It was all second class, second having superseded third in 1956, virtually throughout Europe. After closure, No 1 saw a couple more years work on the West Clare.

Two carriages came from the Tralee & Dingle in 1954, a composite of 1907 and a full brake, formerly a brake third dating from 1890. Steam heat was discussed from time to time but never implemented. Right to the end, foot warmers remained the only form of heating.

In 1925, all railways wholly within the Irish Free State were merged into the Great Southern. After 20 years, this body was amalgamated with road

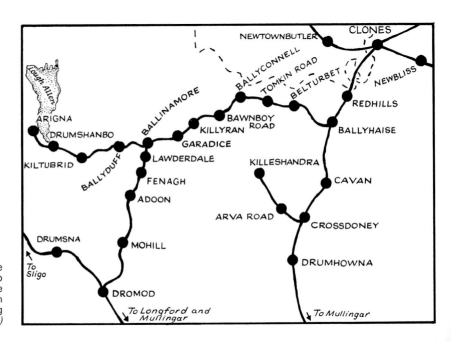

2 – 6 – 0T No 3T crossing the Shannon with an Arigna to Ballinamore mixed. The bridge marks the Roscommon/ Leitrim border, 1½ miles after leaving Arigna. *(Peter Sunderland)*

transport interests to form Coras Iompair Eireann, the Irish Transport Company.

Under the Great Southern, Dromod station facilities and staff were merged with the broad gauge and locomotives were sent away for heavy overhaul. In 1939, the Great Southern proposed to the Transport Tribunal that the Cavan & Leitrim should be closed. Then the Second World War increased the importance of Arigna coal and the railway lasted another 20 years.

It seemed likely that the end would come about 1957 with a plan to sell the entire coal output to a non rail served power station on the shore of Lough Allen. Then the line gained a two year reprieve with a significant contract for the movement of coal to Drogheda and Limerick. Transhipment at Dromod was by hand, as it always had been.

Once this had dried up, closure followed on 31 March 1959.

Cavan & Leitrim Railway Locomotives

No	Name	Type	Built	Arrived	With drawn
1	Isabel	4-4-0T	1887		1948
2	Kathleen	4-4-0T	1887		1959
3	Lady Edith	4-4-0T	1887		1959
4	Violet	4-4-0T	1887		1959
5	Gertrude	4-4-0T	1887		1925
6	May	4-4-0T	1887		1927
7	Olive	4-4-0T	1887		1940
8	Queen Victoria	4-4-0T	1887		1959
9	King Edward	0-6-4T	1904		1934
10		2-4-2T	1900	1934	1959
11		2-4-2T	1900	1934	1939
12		2-4-2T	1900	1934	1959
13		2-4-2T	1900	1934	1951
3T		2-6-0T	1889	1941	1959
4T		2-6-0T	1903	1941	1959
5T		2-6-2T	1892	1950	1959
6T		2-6-0T	1898	1957	1959

4 – 4 – 0T No 8, with nameplate missing, has arrived at Dromod with the 12.45 from Belturbet on 17 May 1924. To celebrate Irish independence, one of the drivers had removed the "Queen Victoria" nameplates and hidden them in Ballinamore Shed Yard. They were found by management and re - affixed. Then another driver unscrewed them whilst the engine was at Drumshanbo, where he buried them permanently.
(LCGB Ken Nunn Collection)

2 – 6 – 0T No 3T racing the competition as it proceeds along the road side with an Arigna to Ballinamore "mixed".
(Peter Sunderland collection)

115

The Cavan & Leitrim ordered eight 4-4-0 tank engines from Robert Stephenson & Co of Newcastle. Two were supposed to have been fitted with forward cabs for working the roadside section but it never happened and Board of Trade rules were quietly forgotten.

The eight locos were named after daughters of the directors apart from No 8 'Queen Victoria'. This engine had its nameplates removed by staff in 1923 to celebrate Irish independence. Withdrawals began in 1925 but four of the class were still in use to the end in 1959.

No 9 'King Edward' was something of a dead loss. A 0-6-4T delivered in 1904, it spread the track and was little used but they couldn't find a buyer for it.

The Great Southern Railway applied the suffix L to the numbers of Cavan & Leitrim engines. P was the corresponding suffix for the Cork, Blackrock & Passage Railway, which closed at the end of 1932. 4P, 5P, 6P and 7P were transferred north, becoming 10L, 11L, 12L and 13L. Built in 1900 by Nielson of Glasgow, two of these survived to the end of the C&L.

The next source of motive power was the Tralee & Dingle. This remained open until 1953 but Nos 3T, 4T and 5T were transferred to the C&L prior to that. 6T remained until closure of the Dingle line. Then it did two years on the West Clare before being replaced by a diesel. It was awaiting an appointment with the scrap man when the coal traffic on the C&L suddenly expanded giving it a reprieve until 1959.

For the first six months, locomotive coal was obtained from England but then they switched to Arigna coal. This was less efficient but the lower cost compensated for higher consumption.

The Extension

During the First World War, the Government decreed that Irish mineral reserves should be exploited and new railways built where necessary.

The 4¼ mile extension from Arigna to Aughabehy was built by the Great Northern Railway as agent for the Government. It opened on 2 June 1920 and was worked by the Cavan & Leitrim. The Extension remained in Government ownership until 1929 when it was leased to the Great Southern for a shilling a year. By that time, the Aughabehy pit was worked out and the line was quickly cut back to Derreenavoggy. The truncated section then remained part of the Cavan & Leitrim until closure in 1959.

Preservation

The spirit of the Cavan & Leitrim never died completely. Some 30 years after closure, revival began at Dromod, where the former narrow gauge station and engine shed have come back to life, along with a short stretch of track. The long term ambition is to reach Mohill.

4 – 4 – 0T No 2L, built 1887, shunting at Ballyconnell in July 1957. *(Peter Sunderland)*

4T entering Arigna with coal empties in July 1957.   *(Peter Sunderland)*

Loading coal at Arigna mine. It was this coal loading facility, on the extension beyond Arigna, which kept the railway open from the mid 1930s until 1959.   *(Peter Sunderland)*

The Clogher Valley Railway

No 3 "Blackwater" arriving at Maguiresbridge with a mixed from Fivemiletown on 14 May 1920. The station building belonged to the Great Northern but dealt with both railways. The narrow gauge continued a short distance behind the camera to the transit shed and CVR engine shed.
(LCGB - Ken Nunn collection)

37 miles in length, this 3ft gauge byway meandered from Tynan, County Armagh, to Maguires Bridge in County Fermanagh. Both termini were shared with the Great Northern Railway.

The line was tortuous with numerous gradients steeper than 1 in 40 and some as much as 1 in 30. Curvature was severe and there were many speed restrictions as low as 4mph. Much of the route was laid along the roadside. Some stretches were laid in the road itself. These included the passage through Fivemiletown where the Summit of the line was reached part way along the Main Street.

William Dargan, the eminent Irish railway engineer, had said that if there was any agricultural district in Ireland that would pay for a railway, it would be the Clogher Valley. At least that is what proponents of the line claimed that he had said – or maybe wished he had said. But the Clogher Valley proved a difficult line to finance and a problem to keep going, even when it had been built. It functioned from 1887 until 1941, closing during the Second World War, which was rather unusual when road transport was so restricted.

The line only became a practical possibility with the Tramways & Public Companies (Ireland) Act, 1883. Earlier attempts had all failed but this legislation afforded a system of local financial assistance, the Baronial guarantee, which was refunded in part by Central Government. The Baronies were a subdivision of the Counties. In each of Fermanagh and Tyrone, there were eight Baronies of which two were to be invited to subsidise the line. County Armagh, with one mile of track did not participate.

The Clogher Valley Tramway Company Limited was incorporated under the Companies Acts, not by private Act of Parliament. Powers to build and operate the line were then conferred by Order in Council on 7 August 1884. The Order was granted only after the promoters had satisfied the Privy Council of Ireland that the venture was desirable in the public interest, that it could not be built without state aid and that forecast revenue was sufficient to make it viable. The promoters were keen to emphasise that they were all local people seeking to improve their area and not outside prospectors looking to profit from the public purse.

The contract to build the line was awarded to McCrea and McFarland of Belfast in April 1885. They had previous experience in Irish railway construction. John McFarland was later to become heavily involved in the Londonderry & Lough Swilly Railway.

With the Baronial guarantee, Clogher Valley shares ought to have enjoyed something approaching gilt-edge status. Indeed, the first batch of shares were issued at a small premium but then demand dried up. The issue had been underwritten by Salter & Son, a merchant bank in Belfast. In 1886, they defaulted on their obligation to buy the unissued shares and the future of the project was placed in severe doubt. With some difficulty, the Company managed to obtain a loan from the Irish Board of Works against the doubtful security of the unissued shares. Work was then able to proceed.

On 11 and 12 April 1887, the line was inspected by Major General Hutchinson of the Board of Trade. Opening was sanctioned for Monday 2 May.

The through journey took a minimum of 2 hours 40 minutes but only one train per day went all the way without significant breaks en route. It is unlikely that any passengers would have travelled from Tynan to Maguires Bridge this way as it would have been quicker on the Great Northern via Clones. *Bradshaw* for April 1910 shows four trains in each direction over all sections of the line but the service is made up of short workings with some trains starting and terminating at Aughnacloy and

Fivemiletown.

Extra trains were timetabled on Ballygawley, Aughnacloy and Fivemiletown Fair Days. *Bradshaw* gives the times without specifying the days involved. *Bradshaw* will have been working from detail sent in by the Clogher Valley Company who probably found it unnecessary to spell out what their customers already knew. The cattle fairs were so much an integral part of the local communities. They were held monthly, Ballygawley, for example, was on the second Friday, occasioning extra traffic both in passengers and in sheep and cattle which, before the line was built, had to be herded along the local highways. More distant Fairs also generated business including Enniskillen and Clones, the latter on the last Thursday of each month. Special trains for these left Clogher Valley stations very early in the morning.

The Clogher Valley opened with 13 passenger coaches. These served, without any additions, for the entire 54 years of operation. On most days, the fleet was more than adequate but was hopelessly insufficient when mass movement of people took place. The busiest day was 13 July when many of the Orangemen had to travel with their bands in improvised cattle wagons and other goods vehicles. Special trains ran to local events such as Tynan Races, also in connection with Great Northern seaside excursions. These could run to either Coast, via Tynan to Newcastle or changing at Maguires Bridge for Bundoran. But never on a Sunday. This embargo was partly religious and partly economic. Not until the 1920s, were Sunday excursions a semi regular feature, usually to Bundoran.

What's in a Name?

The Clogher Valley had been incorporated as a tramway company. This status was useful under the 1883 Act and the line included tramway features, not least where it ran along the public road. However, the Company found the title a disadvantage when dealing with other railways and especially with the Railway Clearing House. It felt it wasn't being taken seriously enough so, in 1894, it obtained Board of Trade sanction for a change of name to The Clogher Valley Railway Company Limited.

Expansion Plans

Having thrown off its status as a struggling tramway and become an impoverished railway, the Clogher Valley saw expansion as the way out of its difficulties. It began to see itself as part of a continuous 3ft gauge system from the Irish Sea to the coalmines served by the Cavan & Leitrim. The Clogher Valley Company supported proposals for new lines, which would link it up with its narrow gauge near neighbours. Support did not extend to any financial commitment. The new lines would be built by raising fresh capital underwritten by Baronial and Government guarantees.

Parliamentary powers were granted in 1901 for 25 miles of line connecting the Bessbrook & Newry at Bessbrook with the Clogher Valley at Tynan. The line would have descended into Tynan, crossing over the Great Northern on the Armagh side of the station, before running in alongside the Clogher Valley. Any through narrow gauge traffic would have had to reverse direction in Tynan Station.

To the West, an Act of 1903 authorised a 22 mile link between the Clogher Valley at Maguires Bridge and the Cavan & Leitrim at Bawnboy Road. It was proposed that once they were all joined up, the narrow gauge Companies would merge.

Nothing came of these ambitious schemes except for a short tunnel at Keady. When the Great Northern built its Armagh to Keady branch, which opened in 1909, it had to provide a tunnel through its embankment just north of Keady Station because the narrow gauge scheme was still on the Statute Book.

Difficult Years

After 1921 the Clogher Valley found itself entirely in Northern Ireland. It suffered some loss of traffic

Augher Station in 1958. This is the platform side, the line running along the roadside. The building is now a coffee shop. Colebrooke Station was similar in design.   *(J C W Halliday)*

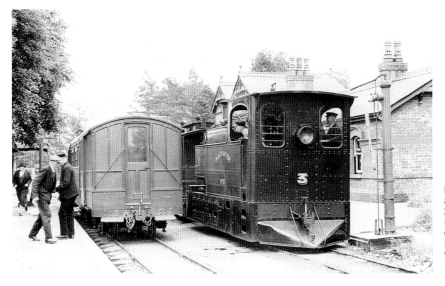

No 3 "Blackwater" has arrived at Fivemiletown from Tynan on 25 June 1937. On the left, No1 railcar and trailer prepare to depart for Tynan. The main building with two gable ends was similar to Clogher.
(H C Casserley)

from its proximity to the Border as there was less movement of cattle and other goods from nearby County Monaghan.

An early action by the new Northern Ireland administration was to commission a Report on the Railways and then ignore its findings. Not that these were very dramatic but the 1922 Commission did recommend that the Clogher Valley be taken over by the Great Northern. This didn't happen.

The Clogher Valley became increasingly unpopular amongst the ratepayers who every year had to stand the full guarantee on the dividend to shareholders. Not that closure of the Railway would diminish their burden. Full payment of the guarantee entitled Tyrone and Fermanagh County Councils to assume ownership but they took no action on this until 1928.

From that year, the Railway was run by a Management Committee appointed by the Government and the two Councils. The Chairman from 1929 was Sir Basil Brooke who later became Prime Minister of Northern Ireland, 1943 to 1963. His Government dismantled the greater part of the railway system in the 1950s. As a youth he had fired engines on the Clogher Valley, unofficially. He wasn't the only person to have done that. It was quite common everywhere.

More important to the development of the line was the appointment to the Management Committee of Henry Forbes, Secretary and Manager of the County Donegal. His influence was present in ordering the Gardner – Walker railcar. In April 1932, Forbes arranged for County Donegal No 4 to visit the Clogher Valley for a trial. Following this, an order was placed with Walker Brothers of Wigan for an articulated car powered by a Gardner diesel engine. This represented a technical advance on railcar technology to date. Clogher Valley No 1 was the prototype for Gardner – Walker railcars elsewhere.

The railcar took over most passenger work allowing an increase in the service with six trains a

day between Tynan and Fivemiletown. It helped reverse a decline in traffic which, by 1932, had fallen to only a third of its pre 1914 level.

Business slumped as a result of the 1933 strike. Although the Clogher Valley men were not involved, there were no Great Northern trains at Maguires Bridge for ten weeks and only a few from Tynan in Belfast direction only. A lot of freight was lost permanently whilst passengers came back only slowly.

Closure
In 1937, the Government agreed terms which would allow closure of the Railway. Since 1930, the Government had been buying shares from holders who were prepared to sell at 60% of their par value. But the shareholders were receiving a guaranteed 5% dividend, representing a yield of 8⅓ % on the Government's offer price. Bank Rate was only 2% for most of the 1930s and 40s. Not surprisingly, many refused to sell and eventually had to be compelled as part of the scheme for closing the Railway and ending the guarantee. The Government accepted responsibility for this, also agreeing to meet the greater part of the cost of road improvements. Eventually, legislation was passed authorising closure at the end of 1941 and subsequent abandonment.

On 31 December, 2-6-2T No 4 worked the last goods train whilst Railcar No 1 handled the last passenger service. This actually left Fivemiletown for Ballygawley at 12.10am on New Year's Day 1942. The Railcar was sold to the County Donegal along with sundry wagons and permanent way materials.

Clogher Valley Motive Power
Six 0-4-2 tank engines were ordered from Sharp, Stewart of Manchester in time for the opening in 1887. Their design incorporated tramway features including skirting, a cowcatcher and a bell. They were fitted with condensers to reduce the emission

No 6 "Erne" arriving at Aughnacloy with the 4.05pm Fivemiletown to Tynan on 18 September 1929.

(H C Casserley)

Clogher Station in 1958. Fivemiletown was of similar construction. *(J C W Halliday)*

of steam and they normally ran cab first so as to give the crew the best possible look out. There were two large front windows between which was mounted a powerful oil headlamp. The one ton coalbunker was on top of the firebox and boiler. Each had a local name. No 4 was the first to be withdrawn in 1929. Three of them lasted to the end, by then little used.

No 1 Caledon
    2 Errigal
    3 Blackwater
    4 Fury
    5 Colebrooke
    6 Erne

In 1910 the Railway ordered a new loco from Hudswell Clarke of Leeds. No 7 "Blessingbourne" was not a success. Withdrawn in 1926, they could not sell it until 1934 when they managed to swap both it and No 1 "Caledon" with a scrap merchant for a new No 4. This was a 2-6-0T built in 1904 for the Castlederg & Victoria Bridge Tramway. The Clogher Valley hesitated about buying it on closure of the Castlederg line in 1933 so it was bought by the scrap man. The subsequent deal involved the Clogher Valley in laying out no cash apart from transport over the G N and modification which was carried out at Aughnacloy. It emerged as a 2-6-2T with larger bunker so that it could travel further from the coaling plant than had been necessary at Castlederg.

No 8 was an Atkinson Walker steam shunter, built in 1928 It was a disaster and the Railway never paid for it. They did, however, sell it for a modest sum to the County Donegal who fitted it with a Gardner diesel engine to become their No 11 "Phoenix".

No 2 in the railcar series was a Gardner – Walker cab and motor unit, identical to No 1 but without the passenger section. Instead it had a small wagon body fixed rigidly behind the cab. It looked a bit like a small road lorry. If the wagon part were removed, the rest of the unit was capable of being exchanged with that of the railcar. This happened permanently in later County Donegal ownership after the railcar suffered a collision.

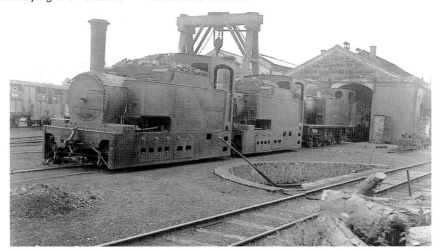

Aughnacloy shed yard in the late 1930s, with two of the surviving Sharp Stewart tram engines in front of No 4, the ex Castlederg 2 – 6 – 2T.
(Martin Bairstow collection)

0 – 4 – 2T No1 "Caledon" a
Fivemiletown in May 1920.
(LCGB – Ken Nunn collection

No 6 "Erne" at Ballygawley with a train for Tynan.
(A W Croughton)

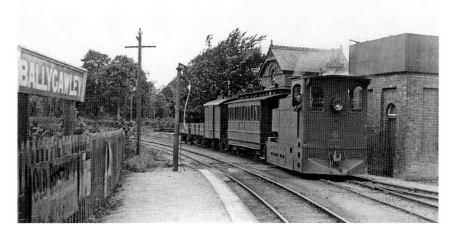

2 – 6 – 2T No 4 ready to leave Tynan about 1936. The Clogher Valley platform curves away from the Great Northern station, the main building of which is off to the left. Closed in 1957, it is still standing.
(A W Croughton)

No 4 in an earlier life at Castlederg, to whom it was supplied by Hudswell Clark as a 2 – 6 – 0T in 1904. The Clogher Valley rebuilt it as a 2 – 6 – 2T with a larger coal bunker for the longer distances it would have to cover. *(A W Croughton)*

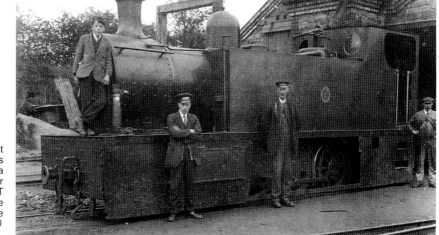

No 1 "Caledon" arriving at Fivemiletown with a mixed from Tynan on 14 May 1924.
(LCGB – Ken Nunn collection)

Railcar No1 and trailer have left Fivemiletown Station and are negotiating the Main Street at the regulation 4mph with the 12.00 to Tynan on 25 June 1937. *(H C Casserley)*

The Castlederg & Victoria Bridge Tramway

Castlederg, County Tyrone, has a population of about 750. From 1884 until 1933, it was connected to Victoria Bridge on the Great Northern by a 3ft gauge roadside tramway.

The first motive power took the form of two steam tram engines, supplied by Kitsons of Leeds. In later years, larger locomotives were obtained from Hudswell Clarke. No 4, a 2-6-0T built in 1904, ended up on the Clogher Valley Railway.

Between 1924 and 1928, use was made of a paraffin powered railcar. After lying derelict, this was sold to the County Donegal Joint Committee in January 1933. In 1928 the Castlederg Company purchased from the NCC one of the Beyer Peacock 2-4-0Ts built for the Ballymena & Larne in 1878.

On Christmas Day 1929, the Castlederg became the first Irish railway to employ diesel power in the form of a 0-6-0 locomotive from Kerr Stuart of Stoke on Trent. During a five month trial, the engine worked some passenger trains but was not successful. The County Donegal turned down the opportunity to buy it.

The line had gradients as steep as 1 in 31. Journey time was 40 minutes for the seven miles with three stops. For much of the railway's life there were four passenger trains each way. The Castlederg was an early casualty of road competition though the timing of closure was determined by the 1933 strike.

0 – 4 – 4T No 5 at Spamont with the 8.15am mixed from Castlederg to Victoria Bridge on 21 May 1924.
(LCGB – Ken Nunn collection)

After closure, all 19 covered wagons were sold to the Clogher Valley Railway, transported to Aughnacloy but never used. Dating mostly from 1884, they occupied a siding there until closure of that Railway in 1942.

(H C Casserley)

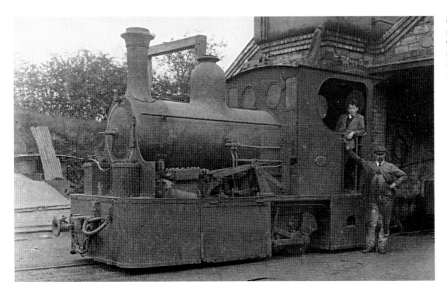

C&VB No 3 was a 0 – 4 – 0T supplied by Kitson in 1891. The wheel guards were on one side only as the tramway kept to one side of the road and there was no turning facility.

(A W Croughton)

Peering out of Castlederg Shed is 2 – 4 – 0T No 6, acquired in 1928 from the LMS Northern Counties Committee. Supplied by Beyer Peacock to the Ballymena & Larne Railway in 1878, it was similar to those still running in the Isle of Man.

(A W Croughton)

Castlederg Station looking towards the buffer stops.

(A W Croughton)

Conclusion

In 1914, there were some 3,500 miles of railway in Ireland. Today, just over 1,200 remain open but they carry many times more passengers than were ever seen on the larger network. Given the low density of population, it is an achievement that Ireland has retained and developed so much. We have to accept that most of the discarded lines really were a lost cause.

There is an important exception to this general statement. A mere glance at the present day map will reveal a most glaring gap north of the line from Dublin to Sligo. In the ancient provinces of Leinster, Munster and Connacht, the railway penetrates all 23 counties. But in the more densely populated Ulster, the system reaches only four out of nine counties.

Amongst the Railways featured in this book, there were once a total of 980 route miles. Now only 117 remain in business, a survival rate of 12%, compared to 35% in Ireland as a whole.

Between 1957 and 1965, railways were banished from Fermanagh, Monaghan, Cavan, Donegal and Tyrone. These areas had their share of hopeless cases which would have failed anywhere in Ireland. But there was another factor at play. It wasn't just the Border, it was the adverse transport policy pursued on the Northern side.

Only in Northern Ireland did the Government describe railways as "Obsolete as the Stagecoach" and promise their removal. This was never threatened in the Republic of Ireland, nor in Great Britain. The "Great Closure" of September 1957, the "tidying up" of adjoining lines and demise of the "Derry Road" created the largest area in the British Isles devoid of railways. The nearest equivalent is the gap on the England/ Scotland border left by closure of the "Waverley Route" in 1969 but it is not impossible that this is going to be remedied.

Had a consistent policy been applied throughout Ireland, there would not be such a huge gap in South Ulster and the Border area. Had Ulster's railways been treated the same as those in Leinster, Munster and Connacht, then it is not two difficult to imagine the "Derry Road", the "Irish North" or Portadown – Armagh – Cavan still plugging this great hole in the railway map.

South of the Border, the Government has adopted a National Spatial Strategy, aimed at preventing the over development of Dublin, concentrating instead on eight "spatial" towns. Seven of these are rail served. There has been vague talk of restoring a rail service to the eighth town, Letterkenny.

he picture which says it all. Great Victoria Street booking office displaying a diagram of destinations eached by the GN. By 1965, only Dublin and Howth still had passenger services.   *(J C W Halliday)*

Opening and Closure Dates

Great Northern Railway (Ireland)
Opened

Belfast - Lisburn	12/08/1839
Lisburn - Lurgan	18/11/1841
Lurgan - Seagoe	31/01/1842
Seagoe - Portadown	12/09/1842
Dublin - Drogheda	24/05/1844
Howth Jn - Howth	30/07/1846
Londonderry - Strabane	19/04/1847
Portadown - Armagh	01/03/1848
Newfoundwell - Dundalk	15/02/1849
Dundalk - Castleblaney	15/02/1849
Newry Dublin Bridge - Warrenpoint	28/05/1849
Drogheda - Navan	15/02/1850
Dundalk - Wellington Inn	31/07/1850
Mullaghglass - Portadown	06/01/1852
Strabane - Newtown Stewart	09/05/1852
Wellington Inn - Mullaghglass	10/06/1852
Newtown Stewart - Omagh	13/09/1852
Navan - Kells	11/06/1853
Omagh - Fintona Town	15/06/1853
Fintona Junction - Dromore Road	16/01/1854
Goraghwood - Newry Edward St	01/03/1854
Castleblaney - Ballybay	17/07/1854
Dromore Road - Enniskillen	19/08/1854
Drogheda - Newfoundwell	05/04/1855
Ballybay - Newbliss	01/07/1855
Portadown - Dungannon	05/04/1858
Armagh - Monaghan	25/05/1858
Newbliss - Lisnaskea	07/07/1858
Lisnaskea - Lisbellow	16/08/1858
Lisbellow - Enniskillen	15/02/1859
Scarva - Banbridge	23/03/1859
Shantonagh Jn - Cootehill	18/10/1860
Newry Edward St - Dublin Bridge	02/09/1861
Dungannon - Omagh	02/09/1861
Clones - Cavan	01/04/1862
Monaghan - Clones	02/03/1863
Kells - Oldcastle	17/05/1863
Knockmore Jn - Banbridge	13/07/1863
Goraghwood - Armagh	25/08/1864
Bundoran Junction - Bundoran	13/06/1866
Knockmore Jn - Antrim	13/11/1871
Dungannon Junction - Cookstown	28/07/1879
Banbridge - Ballyroney	14/12/1880
Ballyhaise - Belturbet	29/06/1885
Inniskeen - Carrickmacross	31/07/1886
Dromin Junction - Ardee	01/08/1896
Sutton - Hill of Howth	17/06/1901
Howth - Hill of Howth	01/08/1901
Ballyroney - Newcastle	24/03/1906
Armagh - Keady	31/05/1909
Keady - Castleblaney	10/11/1910

Closed

Keady - Castleblaney	10/08/1924
Armagh - Keady (passr)	31/12/1931
Goraghwood - Markethill (passr)	31/01/1933
Markethill - Armagh	31/01/1933
Dromin Jn - Ardee (passr)	03/06/1934
Shantonagh Jn - Cootehill (passr)	28/02/1947
Inniskeen - Carrickmacross (passr)	28/02/1947
Shantonagh Jn - Cootehill	30/04/1955
Goraghwood - Markethill	30/04/1955
Scarva - Newcastle	30/04/1955
Dungannon Jn - Cookstown (passr)	16/01/1956
Knockmore Jn - Banbridge	28/04/1956
Bundoran Jn - Bundoran	30/09/1957
Omagh - Clones	30/09/1957
Fintona Junction - Fintona Town	30/09/1957
Portadown - Glasslough	30/09/1957
Armagh - Keady	30/09/1957
Glaslough - Cavan (passr)	13/09/1957
Dundalk - Clones (passr)	13/09/1957
Ballyhaise - Belturbet (passr)	13/09/1957
Drogheda - Oldcastle (passr)	12/04/1958
Glaslough - Monaghan	31/05/1958
Ballyhaise - Belturbet	31/03/1959
Sutton - Hill of Howth - Howth	31/05/1959
Coalisland - Cookstown	30/09/1959
Dundalk - Clones	31/12/1959
Monaghan - Cavan	31/12/1959
Inniskeen - Carrickmacross	31/12/1959
Navan Junction - Oldcastle	31/03/1963
Dungannon Jn - Coalisland	03/01/1965
Goraghwood - Warrenpoint	03/01/1965
Portadown - Londonderry	14/02/1965
Central Jn - Ballymacarret Jn	30/06/1965
Dromin Jn - Ardee	03/11/1975

Sligo, Leitrim & Northern Counties Railway
Opened

Enniskillen - Belcoo (goods)	12/02/1879
Enniskillen - Belcoo (passr)	18/03/1879
Belcoo - Glenfarne	01/01/1880
Glenfarne - Manorhamilton	01/12/1880
Manorhamilton - Collooney	01/09/1881
Collooney - Carrignagat Jn	07/10/1882

Closed

Enniskillen - Carrignagat Jn	30/09/1957

Warrenpoint & Rostrevor Tramway

Opened	01/08/1877
Closed	28/02/1915

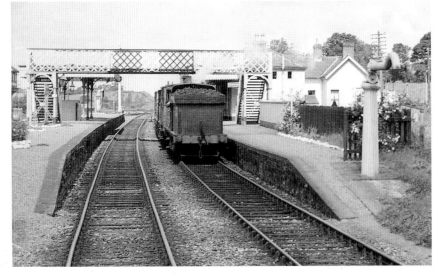

Inniskeen Station, junction f[...]
Carrickmacross, looking west [...]
1958. *(J C W Hallida[...)*